Heavy Ion Collisions at Intermediate Energy

Theoretical Models

Other Related Titles from World Scientific

Multiple Parton Interactions at the LHC
edited by Paolo Bartalini and Jonathan Richard Gaunt
ISBN: 978-981-3227-75-0

ATLAS: A 25-Year Insider Story of the LHC Experiment
by The ATLAS Collaboration
ISBN: 978-981-3271-79-1

Accelerator Physics (4th Edition)
by S Y Lee
ISBN: 978-981-3274-67-9
ISBN: 978-981-3274-78-5 (pbk)

Heavy Ion Collisions at Intermediate Energy

Theoretical Models

Subal Das Gupta

McGill University, Montreal, Canada

Swagata Mallik

Variable Energy Cyclotron Centre, Kolkata, India

Gargi Chaudhuri

Variable Energy Cyclotron Centre, Kolkata, India

World Scientific

NEW JERSEY · LONDON · SINGAPORE · BEIJING · SHANGHAI · HONG KONG · TAIPEI · CHENNAI · TOKYO

Published by

World Scientific Publishing Co. Pte. Ltd.
5 Toh Tuck Link, Singapore 596224
USA office: 27 Warren Street, Suite 401-402, Hackensack, NJ 07601
UK office: 57 Shelton Street, Covent Garden, London WC2H 9HE

Library of Congress Control Number: 2019017469

British Library Cataloguing-in-Publication Data
A catalogue record for this book is available from the British Library.

HEAVY ION COLLISIONS AT INTERMEDIATE ENERGY
Theoretical Models

ISBN 978-981-3277-93-9

For any available supplementary material, please visit
https://www.worldscientific.com/worldscibooks/10.1142/11202#t=suppl

Desk Editor: Ng Kah Fee

Typeset by Stallion Press
Email: enquiries@stallionpress.com

Preface

Heavy ion collisions have been an area of intensive investigation in the last forty years. There continues to be a steady stream of entrants into the field. A short but substantive introductory text in the subject may prove to be useful to many. This book was conceived with this in mind. Heavy ion collision is a huge subject and it is often classified into two separate energy regimes: high energy and intermediate energy. Here we deal with only intermediate energy heavy ion collisions. Somewhat arbitrarily (but conventionally) we restrict here to beam energies from 25 MeV per nucleon to 1 GeV per nucleon.

The task proved to be harder than what we had anticipated. The literature is very huge and often goes into much greater depth than what we could include. We have covered many topics and the omissions are due to (a) our lack of familiarity, (b) technical reasons in putting the subject in a textbook format and (c) we were aiming for a companion textbook for a graduate course in nuclear theory which could be covered in a semester. If the book is sometimes useful as a quick reference to practitioners in the field, we have surpassed our goal. Reading material is so arranged that Chapters 1 to 9 are more thermodynamics based. Chapter 10 onwards deals with more microscopic models. Readers interested mostly in microscopic theories can jump from Chapter 1 to Chapter 10.

We are thankful to many academicians. S.D.G thanks George Bertsch for collaboration during development of microscopic model of intermediate energy heavy ion collision and Aram Mekjian for collaboration during development of macroscopic models (particularly CTM). We are grateful to Betty Tsang, William Lynch, Wolfgang Bauer, Pavel Danielewicz, Michael Mocko and Scott Pratt for welcoming us into the creative environment at NSCL. At McGill S.D.G collaborated with Charles Gale for many years and with Nick de Takacsy before that S.D.G. would like to thank Jean Barrette, Nader Mobed, Jicai Pan, Abhijit Majumder, Suk Joon Lee, Kevin Haglin, H.H. Gan, Gerald Cecil and Champak Baran Das (now deceased) at McGill, Madappa Prakash, Gerd Welke, Tom Kuo and Gerry Brown at Stony Brook. Joe Kapusta's frequent visits to McGill were appreciated.

We thank Wolfgang Trautmann for liberally sharing many experimental data and for encouragement. We thank Francesca Gulminelli and Byron Jennings. Rajat Bhaduri read the first few chapters and recommended some changes. At VECC, we thank Santanu Pal and Dinesh Srivastava for encouraging this project.

Without Juan Gallego's help with the computer and sometimes with physics as well this book would not have seen completion. S.D.G is very appreciative of the hospitality at the theory division of VECC.

Contents

Chapter 1

Introduction

Heavy ion collisions can reveal many properties of nuclear systems otherwise not available in the laboratory. For example, when an energetic heavy ion in the laboratory hits another heavy ion stationary in the laboratory, part of the projectile may be sheared off, and break up into many fragments and continue in predominantly forward direction with some change of velocity. Some of the fragments may be significantly far from the island of stability and properties of this neutron–proton combination may be studied.

The radioactive beam so produced may be made to hit another target to produce even more exotic neutron–proton combination.

If two heavy ions stop each other in head on collisions a blob of nuclear matter at higher-than-normal density will be formed for a short time. The equilibrium density ρ_0 for nuclear systems is about $0.16 \, \mathrm{fm}^{-3}$. At intermediate energy collisions $2/3$ times ρ_0 is reached. The system will then expand often breaking up into many fragments. What are the properties of the dense nuclear systems? How different are they from the properties of nuclear systems at normal density? What is the compressibility coefficient of nuclear matter? What experiments can give a measure of the compressibility of nuclear systems?

In heavy ion collisions new particles can be produced and their production cross-section measured. At intermediate energy the only significant particles produced are pions. By "intermediate energy" we mean beam energy between $25 \, \mathrm{MeV/nucleon}$ to $1 \, \mathrm{GeV/nucleon}$.

In low energy heavy ion collisions (below 25 MeV/nucleon), the reaction mechanism is dominated by mean field as collisions are Pauli blocked and the main reaction channels are fission, particle evaporation etc. On the other side, for high energy reaction (above 1 GeV/nucleon) the mechanism is dominated by collision and the effect of mean field is less important. But at intermediate energy domain there is a competition between mean field and collision. Is the pion production cross-section sensitive to nuclear equation of state, to nuclear incompressibility? In later sections we will deal with this question. We will find that measurement of "flow angles" is a better probe for incompressibility.

Do we expect to see liquid–gas phase transition in nuclear matter? To study this we need to create large chunks of matter. Heavy ion collision gives us the possibility of producing (even though only for a short time) nuclear systems larger than what is readily available in the lab. Many features of nuclear ground states are well described by the liquid drop model so one might expect to see liquid–gas phase transition in large nuclear system at non-zero temperature. Heavy ion collisions provide us the best chances of observing signatures of phase transition because not only we can create non-normal density but also create this at optimal temperatures. But exploration and interpretation of data will not be easy. Phase transition occurs in infinite systems but in nuclear physics Coulomb repulsion limits the size to which nuclei can grow. However we believe we can recognise the distortion brought in by finite size. We will look at a few examples.

As is well known, a different kind of phase transition occurs at very high energy heavy ion collisions. Nucleons lose their individual identity and merge into a quark–gluon plasma. This is a very active field of research. We will be working in a much lower energy domain where nucleons retain their identity and features derivable from non-relativistic quantum mechanics are retained.

A great deal of data has accumulated in heavy ion collisions at intermediate energy. The fragments emerging from these span a very wide range of the nuclear periodic table. In general the fits obtained with theoretical models are very good. Understanding the

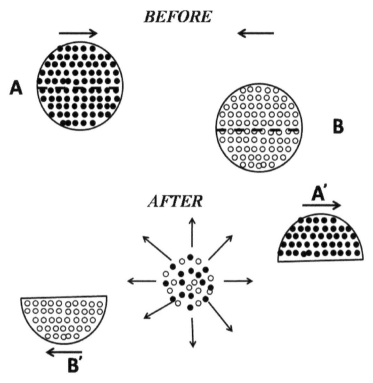

Fig. 1.1: Pictorial view of heavy ion reaction in the centre-of-mass frame. A certain part of the projectile B overlaps with a certain part of the target A. They are participants. In the lower figure we denote by B' the projectile spectator and by A' the target spectator.

data required a lot of statistical mechanics. Microcanonical, canonical and grand canonical ensembles have been used. This has been a fertile ground for comparisons of results with different ensembles.

Consider projectile ion B hitting target ion A stationary in the lab (see Fig. 1.1). The impact parameter is b. In straight line trajectory some part of B will not encounter any part of A. If the incident energy is high enough this part will shear off the rest of B and will continue predominantly in the forward direction. We label this the projectile spectator. We can expect the projectile spectator to have some excitation because this does not have the "best" shape and because of migration of some nucleons across the dividing plane.

Similarly we can expect target spectator. The target spectator will be basically stationary in the lab. Then there are regions $V_B(b)$ of ion B and $V_A(b)$ of ion A where the two are in each other's way. There will be hard nucleon–nucleon collisions as nucleons in $V_B(b)$ try to traverse $V_A(b)$. The nucleons that belonged to $V_B(b)$ and $V_A(b)$ are called participants. We have indicated that volumes V_A, V_B are dependent on impact parameter. So are the volumes of projectile and target spectators. They are sometimes called PLF (projectile like fragment) and TLF (target like fragment).

For many calculations one needs the values of the volumes of PLF, TLF and participating regions $V_B(b)$ and $V_A(b)$. Formulae for these are given in Appendix A.

The concept of temperature can be brought in by postulating that because of nucleon–nucleon collisions nucleons in V_A and V_B (the participants) will fuse. It is like two blobs of clay hit one another and became one. Let us put in nuclear physics numbers of relevance to us. Suppose the beam energy is 100 MeV/A and we have $V_B(b)$ with 40 nucleons hitting $V_A(b)$ containing 60 nucleons. Going to the centre of mass (cm) we find the energy of relative motion is 2400 MeV to be shared by 100 nucleons. The binding energy per nucleon is about 8 MeV per nucleon which is well below the 24 MeV per nucleon brought in by the beam. This implies all the nucleons in V_B and V_A will be liberated. The average energy per nucleon is 16 MeV. Thus we have a "fireball" at temperature 10.67 MeV if we assume that what emerges from this hot volume is just a classical gas of neutrons and protons. If this is what happens then knowing the excitation energy brought in by the beam (usually known) is enough to specify the temperature. But experimentally not only bare nucleons but also composites like deuteron, triton, alpha and many other species come out. We expect initially the participating zone will be compressed then nuclear matter will expand. As it expands nuclear matter will redistribute into monomers and composites. We have to find out how this redistribution is to be calculated. What is also significant is that in this situation knowing the excitation energy does not immediately give us the temperature. In fact the reverse route is taken. If we guess a temperature it is easier to calculate the energy.

We will in later chapters develop formulae which will allow us to calculate the redistribution of the nucleons in the participant zone assuming a canonical thermodynamic model (CTM). This has become well known in last twenty years. Before that one used the grand canonical ensemble to estimate productions of monomers and composites assuming thermal and chemical equilibrium. We will look into this now.

We assume we are in the participant zone, which was compressed initially, then expanded. The nucleons got distributed into different species. We need two indices to characterise a species. The first index will refer to the neutron number and the second the proton number. Thus a neutron is 1,0; a proton is 0,1; a deuteron is 1,1; a triton is 2,1; etc. The expanded volume is large enough so that nuclear interactions between final products can be neglected. Consider one species first. The partition function of one particle of this species be denoted by $\omega_{i,j}$. If the species is not a composite $\omega_{i,j}$ is $\frac{V}{h^3}(2\pi mT)^{3/2}$ (arising from $\iint \frac{d^3r d^3p}{h^3} \exp(-p^2/(2mT))$). If the species is a composite the value of $\omega_{i,j}$ is multiplied by an additional factor to include the binding energy and excited states of the composite. We will deal with this aspect in later chapters. In the expression for $\omega_{i,j}$, V is the volume in which the particles move, m is the mass of the species. T here has the Boltzmann factor absorbed and has the dimension of energy. T will be in MeV. We will also use $\beta = 1/T$.

If $\omega_{i,j}$ is the partition function of one particle of species i, j, then the canonical partition function $Q_{n_{i,j}}$ for $n_{i,j}$ particles of species i, j is $Q_{n_{i,j}} = \frac{(\omega_{i,j})^{n_{i,j}}}{n_{i,j}!}$. Grand canonical ensemble allows for the whole range of values of n. The grand partition function is given by

$$Z_{gr}(i,j) = \sum_0^\infty \exp(\beta\mu_{i,j}n_{i,j})\frac{\omega_{i,j}^{n_{i,j}}}{n_{i,j}!} \tag{1.1}$$

From Eq. (1.1) the following easily emerges. $\log Z_{gr}(i,j) = e^{\beta\mu_{i,j}}\omega_{i,j}$; average $n_{i,j} = \langle n_{i,j}\rangle = \frac{\partial \log Z_{gr}(i,j)}{\partial \beta\mu_{i,j}} = e^{\beta\mu_{i,j}}\omega_{i,j}$. Including all the species now, we have

$$\log Z_{gr} = \sum_{i,j} e^{\beta\mu_{i,j}}\omega_{i,j} \tag{1.2}$$

There are as many μ's as species. We now assume that for our system there is thermal as well as chemical equilibrium. This imposes relationships between chemical potentials of different species [1, 2]. For example $p + n + N$ going to $d + N$ or *vice versa* implies that at equilibrium $\mu_d = \mu_n + \mu_p$; $\mu_t = 2\mu_n + \mu_p$; $\mu_{3\mathrm{He}} = 2\mu_p + \mu_n$ etc., and all chemical potentials can be expressed in terms of neutron chemical potential μ_n and proton chemical potential μ_p.

Going back to the problem at hand, we know how many neutrons N and protons Z are participants. We need

$$N = \sum_{i,j} i \times \exp(i\beta\mu_n + j\beta\mu_p)\omega_{i,j} \qquad (1.3)$$

$$Z = \sum_{i,j} j \times \exp(i\beta\mu_n + j\beta\mu_p)\omega_{i,j} \qquad (1.4)$$

For a given temperature, the above two equations determine chemical potentials μ_n and μ_p. The average values of occupations $\langle n_{i,j} \rangle$ of all the species i, j can now be found.

In the late seventies and early eighties the grand canonical ensemble was used many times. For an exhaustive calculation, see [3]. More recently the grand canonical model was used to study phase transition [4, 5]. In this book we refer to the grand canonical model as GCM. A common alternative name is macrocanonical model.

Although we introduced the grand canonical model here keeping the participant zone in mind, the model can also be applied to PLF or TLF since they get excited when formed.

Chapter 2

A Simple Model for Nuclear Multifragmentation

2.1 A simple canonical thermodynamic model

We determined, using the postulate of thermal and chemical equilibrium, the distribution of monomers and composites that result from an assembly of nucleons at a finite temperature. The grand canonical ensemble was used. The grand canonical ensemble does not conserve the total number of particles. Using chemical potentials μ_n and μ_p we can ensure that the average numbers of neutrons and protons equal N and P present in the fireball. But there are fluctuations. If N and P are very large numbers these fluctuations do not matter. But in heavy ion collisions we often have to deal with situations where only 40 or 50 nucleons are involved. In such cases fluctuations which are present in the theory but not in reality can predict quite wrong results. The canonical model conserves particle numbers but it was not known how to use the canonical model for this problem. That changed in 1995 [6].

Composites have two indices i = neutron number of the composite and j = proton number of the composite. Let $n_{i,j}$ denote the number of composites with i neutrons and j protons. The allowed exit channels have to satisfy $\sum_{i,j} i \times n_{i,j} = N$ and $\sum_{i,j} j \times n_{i,j} = P$. It is awkward to impose this in a calculation and there are far too many choices of $n_{i,j}$ that satisfy this. The algebra that enables us to do this will follow, but we first do some simpler model calculations in this chapter. We consider one kind of particles. The reason for doing this is (a) the algebra is simpler, (b) this allows us to consider systems of very large number of particles, and (c) clean evidence of phase transition already appears. We have species labelled by i. Here $i = 1$ would mean a monomer (one nucleon); $i = 2$ means a dimer, a composite made of two nucleons; $i = 3$ means a trimer, a bound system of three nucleons; etc. The partition function of one particle of species i is denoted by ω_i and the partition function of n_i particles of type i is $\frac{(\omega_i)^{n_i}}{n_i!}$. The partition function of our total system with A nucleons which has many species i is

$$Q_A = \sum \prod \frac{(\omega_i)^{n_i}}{n_i!} \tag{2.1}$$

Only partitions that satisfy $\sum i n_i = A$ are allowed. The number of partitions that satisfy this is enormous. We call each partition that satisfies this a channel. The probability of a channel $P(\vec{n}) = P(n_1, n_2, n_3, \dots)$ is

$$P(\vec{n}) = \frac{1}{Q_A} \prod \frac{(\omega_i)^{n_i}}{n_i!} \tag{2.2}$$

The average number of composites which have i bound nucleons is easily deduced from the above equation.

$$\langle n_i \rangle = \omega_i \frac{Q_{A-i}}{Q_A} \tag{2.3}$$

Since $\sum i \langle n_i \rangle = A$ we get a recursion relation [6]

$$Q_A = \sum_i^A i \omega_i Q_{A-i} \tag{2.4}$$

Let us write down the first few of these. Starting with $Q_0 = 1$ we have $Q_1 = \omega_1$, then $Q_2 = (1/2)\omega_1^2 + \omega_2$, then $Q_3 = (1/6)\omega_1^3 + \omega_2\omega_1 + \omega_3$, etc. Even for $A = 3000$ the above equation takes negligible time to compute. Once we know the canonical partition function Q_A we can calculate observables of interest.

The quantity ω_i is a product of two parts:

$$\omega_i = \frac{V}{h^3} (2\pi m T)^{3/2} i^{3/2} \times z_i(int) \tag{2.5}$$

We have encountered the first part of the right hand side before. It arises from the motion of the CM of the composite i in a volume V; $z_i(int)$ is the intrinsic partition function of the composite. This contains all the nuclear physics and is responsible for all the interesting phenomena like phase transition etc. To get an expression for $z_i(int)$ we use the thermodynamic identity $z = \exp(-F/T)$ where F is the free energy of internal degrees of freedom of the composite. We have $F = E - TS = E_0 + E_{exc} - TS$. In Fermi-gas model the expression for E_{exc} at intermediate energy [7] is iT^2/ϵ_0 and TS is $TS = T \times 2iT/\epsilon_0$ where ϵ_0 is a constant. This gives $F = E_0 - T^2 i/\epsilon_0$. We take $E_0 = -W_0 i + \sigma(T)i^{2/3}$ where $W_0 \approx 16\,\mathrm{MeV}$ is the so-called volume energy term and $\sigma(T)$ is temperature-dependent surface tension term. This gives

$$z_i(int) = \exp[(W_0 i - \sigma(T)i^{2/3} + T^2 i/\epsilon_0)/T] \qquad (2.6)$$

The value of ϵ_0 is taken to be $16\,\mathrm{MeV}$. For surface term we use $\sigma(T) = \sigma_0[(T_c^2 - T^2)/(T_c^2 + T^2)]^{5/4}$ with $\sigma_0 = 18\,\mathrm{MeV}$ and $T_c = 18\,\mathrm{MeV}$ [8]. The volume V in the above equation is not the freeze-out volume but somewhat less to allow for volumes of composites. In Van der Waal's spirit V is taken to be $V_{\text{freeze}} - A/\rho_0$; ρ_0 is normal nuclear density. We denote A/V_{freeze} by ρ. Using $V = V_{\text{freeze}} - V_0$ is an acceptable approximation if V_{freeze} is large compared to A/ρ_0. More sophisticated approximations have not been used in practical calculations. Some typical results that will be presented are not sensitive to moderate but non-negligible changes to V.

We need to write down the energy and pressure of the system. We expect $E = \sum \langle n_i \rangle E_i$ and $P = \sum \langle n_i \rangle T/V$ which is just the law of partial pressure. We can directly verify these identities. For example, using $E_i = T^2 \partial \log \omega_i/\partial T$, we have

$$E_i = \frac{3}{2}T + i(-W_0 + T^2/\epsilon_0) + \sigma(T)i^{2/3} - T[i\partial\sigma(T)/\partial T]i^{2/3} \quad (2.7)$$

Of these the first term is the kinetic energy of the centre of mass. The $\partial\sigma(T)/\partial T$ term has little effect. For the total energy we need to calculate $T^2(1/Q_A)\partial Q_A/\partial T$. Using Eq. (2.1) and some algebra we get the anticipated result $E = \sum \langle n_i \rangle E_i$.

There are many interesting quantities one can calculate with the canonical model. Here we will calculate the average value of the largest cluster in the ensemble as that will appear as an order parameter for phase transition. If there are A particles, there is one channel of A monomers which is represented by $\omega_1^A/A!$. The highest cluster in this channel is a monomer. Then there is a channel represented by $(\omega_1^n/n!)(\omega_2)^{A/2-n/2}/(A/2-n/2)!$. In this channel the heaviest cluster is a dimer. Let us label by $Q_A(\omega_1, \omega_2, \omega_3, \ldots, \omega_k, 0, 0, 0, \ldots)$ the A-particle partition function where all ω's for composites with nucleon numbers greater than k have been set to 0. In this ensemble, the largest cluster will span from monomer up to species k. Let us also denote by $Q_A(\omega_1, \omega_2, \omega_3, \ldots, \omega_{k-1}, 0, 0, 0, \ldots)$, the partition function of an ensemble where the largest non-zero ω is that of $k-1$. In this ensemble all the previous channels are included except the ones where the largest cluster has k nucleons. Denote by

$$\Delta Q_A(k) = Q_A(\omega_1, \omega_2, \omega_3, \ldots, \omega_k, 0, 0, \ldots)$$
$$-Q_A(\omega_1, \omega_2, \omega_3, \ldots, \omega_{k-1}, 0, 0, 0 \ldots) \qquad (2.8)$$

then the probability $P_{max}(k)$ of obtaining species k as the largest cluster is given by

$$P_{max}(k) = \frac{\Delta Q_A(k)}{Q_A(\omega_1, \omega_2, \omega_3, \ldots, \omega_A)} \qquad (2.9)$$

The average value of the largest cluster is given by $\langle k_{max} \rangle = \sum P_{max}(k) \times k$. The more interesting quantity is $\langle k_{max} \rangle / A$. The ratio limits are ≈ 0 and 1. The lower values would indicate a gaseous state whereas high value would imply there is a large cluster which would represent a liquid.

There are many other physical quantities one can calculate [9] but let us now investigate if we can draw any conclusions about onsetting of a phase transition.

2.2   Indicating the onset of phase transition

In Fig. 2.1 we have plotted for a system of 200 particles $\langle n_i \rangle$ as a function of i at 3 different temperatures. For this calculation we assumed $V_{\text{freeze-out}} = 3.7 V_0$ where V_0 is normal nuclear volume of a

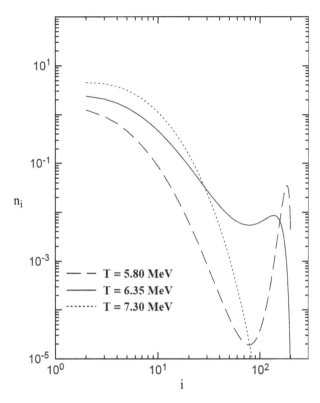

Fig. 2.1: Mass distribution of fragments produced from a system of 200 particles at three different temperatures: 5.8 MeV (dashed line), 6.35 MeV (solid line) and 7.30 MeV (dotted line). Calculations are done at constant freeze-out volume $3.7V_0$.

nucleus of 200 particles. The V that appears in CM motion is then $2.7V_0$. The numbers $\langle n_i \rangle$ in Fig. 2.1 if multiplied by some constant area σ would be the inclusive cross-sections of composites of mass i. At the lowest temperature of 5.80 MeV shown in the figure, $\langle n_i \rangle$ first drops as a function of i, reaches a minimum and then rises, reaches a maximum near the limit and then drops to zero. There will be a heavy fragment in virtually every channel. That heavy fragment is considered to be liquid. As the temperature rises the height of the second maximum drops. It has almost vanished at $T = 6.35$ MeV marking this temperature as the "boiling temperature". Above this temperature ($T = 7.30$ MeV) there are no heavy particles, only light particles. Two points to note. Surface tension is crucial for a large blob to appear. At equilibrium, free energy $F = E - TS$ minimises. If the temperature is small the energy term dominates. The volume term $-W_0 A$ does not distinguish between the energies of $i = A$ and say, 2 specimens of $i = A/2$ but the surface term makes $i = A$ preferable. The entropy term $-TS$ drives the system towards breakup. The temperature T at which the second maximum disappears is dependent upon the number of particles in the system but not sensitively so.

Figure 2.2 displays several signatures of first-order phase transition. The free energy is given by $F = -T \log Q_A$ and we know the value of Q_A. F/A is plotted vs T for $A = 200$ and $A = 2000$ in the top part of Fig. 2.2. A break in the derivative of F/A is seen to occur at $T \approx 6.35\,\text{MeV}$ for 200 particles and at $\approx 7.15\,\text{MeV}$ for 2000 particles. This is a signature of first-order phase transition [10, 11]. The hallmark of a first-order phase transition is the appearance of a maximum of specific heat at constant volume C_V at the phase transition temperature. This is reproduced in the calculation. We will find this to be a very useful signature later. The lowest diagram in Fig. 2.2 is very interesting too. It shows that below the phase transition temperature the average value of the largest cluster is almost the entire system whereas above, the system almost immediately breaks up into small particles. The liquid has transformed into gas.

2.3 The same model in the grand canonical ensemble

The same model (one kind of particles) can be easily implemented in the grand canonical ensemble. Let us consider the case of 200 particles. We restrict the species to composites of 200 nucleons or less. The species would have chemical potentials $\mu, 2\mu, 3\mu, 4\mu, \ldots$ etc. Given a temperature, one just has to adjust one constant μ so that $\sum_1^{200} \exp(\beta i \mu)\omega_i = 200$. The value of ω_i can be read off Eqs. (2.5) and (2.6). It is of interest to compute the energy E. We first outline the steps to calculate the contribution by one species i. We write $\ln z_{gr}(i) = e^{\beta i \mu}\omega_i = e^{i\lambda}\omega_i$ where instead of treating

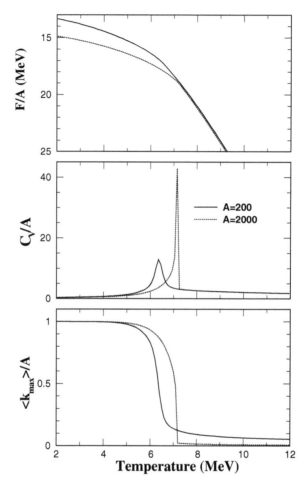

Fig. 2.2: The Helmholtz free energy per particle (upper panel), the specific heat at constant volume C_V/A (middle panel) and the size of the largest cluster (lower panel) as a function of temperature for two fragmenting systems of 200 (solid lines) and 2000 (dotted lines) particles.

β and μ as independent variables we use β and λ as independent variables. The contribution to energy E from i is $-\frac{\partial}{\partial\beta}\ln z_{gr}(i) = -e^{i\lambda}\frac{\partial}{\partial\beta}\omega_i = -e^{i\lambda}\omega_i\frac{1}{\omega_i}\frac{\partial}{\partial\beta}\omega_i = -e^{i\lambda}\omega_i\frac{\partial}{\partial\beta}\ln\omega_i = \langle n_i\rangle E_i$ where E_i is given by Eq. (2.7) and $\langle n_i\rangle$, the grand canonical value of the average number of species i. Total E is the sum of the contributions of all i's.

Chapter 3

Towards Realistic Calculations

3.1 Realistic two-component canonical thermodynamic model

We now extend the canonical thermodynamic model (CTM) to a system of two kinds of particles: protons and neutrons [12, 13]. Extension to three kinds of particles has also been done [14] but we will not treat that here. With two kinds of particles we can label composites with two indices i, j with the first index i giving the neutron number and the second index j giving the proton number. The partition function would be [12]

$$Q_{N,Z} = \sum \prod \frac{\omega_{i,j}^{n_{i,j}}}{n_{i,j}!} \qquad (3.1)$$

The partitioning into composites has to be such that $Z = \sum j \times n_{i,j}$ and $N = \sum i \times n_{i,j}$. Sometimes it is useful to label the composite with mass number a and proton number j. In such cases we will use the first index a for mass number and the second index j for proton number. We will then write

$$Q_{A,Z} = \sum \prod \frac{\omega_{a,j}^{n_{a,j}}}{n_{a,j}!} \qquad (3.2)$$

The sum rules will be $A = \sum a \times n_{a,j}$ and $Z = \sum j \times n_{a,j}$. The partition function $Q_{A,Z}$ has the same value as $Q_{N,Z}$ although the recursion relations for the two look different. For $Q_{N,Z}$ use $\langle n_{i,j} \rangle = \frac{1}{Q_{N,Z}} \sum \prod n_{i,j} \frac{\omega_{i,j}^{n_{i,j}}}{n_{i,j}!} = \omega_{i,j} Q_{N-i,Z-j}$. Using $N = \sum i \times \langle n_{i,j} \rangle$ we have

15

the recursion relation

$$Q_{N,Z} = \frac{1}{N} \sum_{i,j} i\omega_{i,j} Q_{N-i,Z-j} \qquad (3.3)$$

Similarly we can derive

$$Q_{A,J} = \frac{1}{A} \sum_{a,j} a\omega_{a,j} Q_{A-a,J-j} \qquad (3.4)$$

We need to specify $\omega_{i,j}$ with i neutrons and j protons in the composite. As in previous chapter we have

$$\omega_{i,j} = \frac{V}{h^3} (2\pi mT)^{3/2} a^{3/2} \times z_{i,j}(int) \qquad (3.5)$$

Here $a = i + j$ is the total mass of the composite. The proton and the neutron are the fundamental building blocks hence $z_{1,0}(int) = z_{0,1}(int) = 2$ to incorporate spin degeneracy. A reasonable enumeration of $z_{i,j}(int)$ for composites is the following. For deuteron, triton, ^{3}He and ^{4}He, $z_{i,j}(int) = (2S_{i,j} + 1)\exp(-E_{i,j}(gr)/T)$. Here $E_{i,j}(gr)$ is the ground state energy and $2S_{i,j} + 1$ is the ground state spin degeneracy. Excited states of these low mass nuclei are not included. For mass number $a = 5$ and higher, a practical choice is the liquid drop formula ($a = i + j$)

$$z_{i,j}(int) = \exp\frac{1}{T}\left[W_0 a - \sigma(T)a^{2/3} - \kappa\frac{j^2}{a^{1/3}} - c_s\frac{(i-j)^2}{a} + \frac{T^2 a}{\epsilon_0}\right]$$
$$(3.6)$$

Compare these equations with Eqs. (2.5) and (2.6). Compared to Eq. (2.6) the new terms introduced are the Coulomb self-energy terms ($\kappa = 0.72\,\text{MeV}$) and symmetry energy term ($c_s = 23.5\,\text{MeV}$). We also need to specify apart from the ones already mentioned which other composites with neutron number i and proton number j are included in the partition function. We include all nuclei in a ridge centering the line of stability. The liquid drop formula provides drip lines. In some calculations all nuclei within drip lines are included. Nuclei far from the line of stability make negligible contributions.

The Coulomb interaction is long range. Some effects of the Coulomb interaction between different composites can be included in an approximation called the Wigner–Seitz approximation [8]. We assume, as usual, that the break up into different composites occurs at a radius R_c which is greater than the normal radius R_0. Considering this as a process in which a uniform dilute charge distribution within radius R_c collapses successively into denser blobs of proper radius $R_{i,j}$ we write the Coulomb energy as

$$E_c = \frac{3}{5}\frac{Z^2 e^2}{R_c} + \sum_{i,j} \frac{3j^2 e^2}{5R_{i,j}}\left(1 - \frac{R_0}{R_c}\right) \qquad (3.7)$$

It is seen that the expression is corect at the two extreme limits: very large freeze-out volume ($R_c \to \infty$) or if the freeze-out volume is the normal nuclear volume so that there is just one nucleus with the proper radius.

In the model we are pursuing, the term $\frac{3}{5}Z^2 e^2/R_c$ is of no significance since the freeze-out volume is constant. In mean-field sense then one would just replace the Coulomb term $\kappa\frac{j^2}{a^{1/3}}$ by $\kappa\frac{j^2}{a^{1/3}} \times (1.0 - (\rho/\rho_0)^{1/3})$.

Lastly, we write down equations that are needed for computing total energy at a given temperature. Let us see how Eq. (2.7) will change. It is usual to omit the negligible $\partial\sigma(T)/\partial T$ term. Instead of E_i we have $E_{i,j}$ where i is the neutron number of the composite and j is the proton number. Call $a = i + j$ the mass number of the composite. Then for $a \geq 5$,

$$E_{i,j} = \frac{3}{2}T - W_0 a + \sigma a^{2/3} + \kappa\frac{j^2}{a^{1/3}} + c_s\frac{(i-j)^2}{a} + \frac{T^2 a}{\epsilon_0} \qquad (3.8)$$

To approximately account for long range Coulomb interaction we can replace the constant κ by $\kappa \times (1.0 - (\rho/\rho_0)^{1/3})$. For $a < 5$, use for $E_{i,j}$ ground state energy plus $(3/2)T$. The total energy is $E = \sum_{i,j}\langle n_{i,j}\rangle E_{i,j}$ where $\langle n_{i,j}\rangle$ is taken from CTM or GCM.

Calculation for caloric curve (E^* versus T) in the CTM was done in [12, 15]. A plateau in the caloric curve is found around 5 MeV which is in accordance with experimental finding. An interesting point in the calculation is the following observation. Without the Coulomb interaction the height in the peak of the specific heat increases with fragmenting system mass (see Fig. 2.2). With

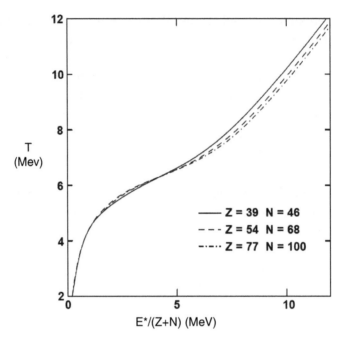

Fig. 3.1:   Nuclear caloric curves for three different nuclei.

Coulomb interaction the height is reduced and the dependence on fragmenting system mass nearly disappears. This growth in size is compensated by the growth in Coulomb repulsion. This means the caloric curve is approximately universal, i.e. it does not depend strongly on the specific system which is disintegrating. The caloric curves computed for three disintegrating systems are shown in Fig. 3.1.

3.2   A comparative study of CTM with other statistical models

We are in a position to compare in the nuclear case, the predictions for yields of composites calculated with CTM and GCM. We compare two cases: $A = 200$, $Z = 80$ (Fig. 3.2) and $A = 50$ and $Z = 25$ (Fig. 3.3). While CTM can exactly reproduce these numbers, with GCM we have not just $A = 200$, $Z = 80$ but also other A, Z's both

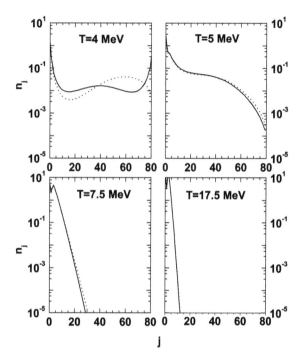

Fig. 3.2: Charge distribution of fragments produced from a fragmenting system of mass number 200 and atomic number 80 obtained from CTM (solid lines) and GCM (dashed lines).

above and below 200 and 80. What is guaranteed is that the average values are 200 and 80. Similarly for the case of $A = 50, Z = 25$. Since there are too many composites we compare isotope yields (yields of the same Z are added up and then compared). The CTM and GCM predictions are quite close at high temperatures, but at low temperatures ($\sim 4\,\text{MeV}$) in spite of 200 being a large number, the GCM predictions are significantly different. One encounters such temperatures in intermediate energy heavy ion collisions, thus one would conclude one should not use GCM at intermediate energies. In fact CTM is numerically easy to solve, sometimes easier than GCM which requires an iterative procedure to arrive at the correct values of μ_n and μ_p. That may require quite precise manipulations. On the other hand, the building of the partition function $Q_{N,Z}$ from bottom up using the recursion relation requires no trial and error.

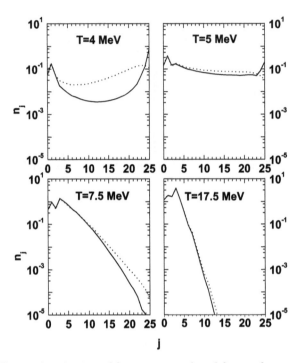

Fig. 3.3: Charge distribution of fragments produced from a fragmenting system of mass number 50 and atomic number 25 obtained from CTM (solid lines) and GCM (dashed lines).

The most dominant model used for data fitting in the last twenty years has been the Copenhagen model (also called SMM or SMFM, statistical multifragmentation model) and we will make some comparison of the yields from SMM with yields from CTM. The sum in Eqs. (3.1) or (3.2) is over many, many possible partitions each of which satisfies $\sum_{a,j} a n_{a,j} = A$ and $\sum_{a,j} j n_{a,j} = Z$. (We are using a, j instead of i, j to be in conformity with [16, 17]). For calculating the average yield of a composite our prescription for CTM avoided having to go through the enumeration of partitions. The Copenhagen model goes through the partitions using Monte Carlo sampling. To give an impression of the enormity of numbers, the number of partitions for $A = 10, Z = 5$ is 339; for $A = 30, Z = 15$, this jumps to 5,236,586. Some manipulations are necessary to avoid building partitions which will have no or insignificant contributions.

Details of how this is done is beyond the scope of this book. Long algebraic manipulations are necessary and [8, 16, 17] deal with this. However, the following observation is useful. Let the number of partitions for a system with mass number A and charge number Z be denoted by $N_p(A, Z)$. This is the huge number we were referring to. Each partition is also characterised by multiplicity $M = \sum_{a,j} n_{a,j}$. (Remember $A = \sum_{a,j} a n_{a,j}$ and $Z = \sum_{a,j} j n_{a,j}$.) We can write $N_p(A, Z) = \sum_M N_p(A, Z, M)$ and at a given temperature only partitions in a small range of M contribute to the weight. For the case of $A = 100$ and $Z = 44$, Fig. 3.1 in [16] shows that at temperature of $2\,\text{MeV}$, partitions with $M = 1, 2$ need to be considered, whereas at temperature $6\,\text{MeV}$ partitions with M between 2 and 7 need to be considered.

For comparison of yields of composites in CTM and SMM several (A, Z)'s at different temperatures were considered in [18]. We discuss these cases now. At low excitation energy, SMM code uses a microcanonical treatment, but taking into account a limited number of disintegration channels. As a rule, only partitions with total fragment multiplicity $M \leq 3$ are considered. For $A = 73$ and $Z = 32$ with excitation $1\,\text{MeV/nucleon}$ (corresponding to $T = 3.13\,\text{MeV}$ for CTM) the yields from SMM and CTM were compared in Fig. 3.1 of [18]. Except where the yields are very small, the agreement is excellent. In Fig. 3.2 of [18] the CTM and SMM models are compared at temperatures of 5 and $8\,\text{MeV}$ for three systems. One system is the same as before ($A = 73$ and $Z = 32$), another system is twice this size and the third one is more neutron rich ($A = 146$ and $Z = 56$). For these the SMM partitions were generated by Monte Carlo sampling from the GCM distribution. The agreement is very good. It should be remembered that the CTM results are exact; the results of SMM are not but obviously the Monte Carlo sampling is very good. The freeze-out volume in all these examples were six times normal nuclear volume. The CTM calculations are easy compared to SMM calculations but SMM can allow different freeze-out volumes for different channels whereas the CTM prescription used here would not allow that. We have used CTM here to calculate only average values of yields but many other results can be obtained [19, 20]. For example, multiplicity distributions of intermediate mass fragments (intermediate mass means mass number between 6 and 40) which are subjects of several investigations [9,21,22] can be calculated exactly.

We have described CTM and GCM and introduced SMM but have not discussed the microcanonical model. This is discussed in [23, 24]. Implementation of the microcanonical model to fit most experimental data is very difficult.

In the next section we give a method of calculating effects of evaporation. This is a very old problem and there are many different methods of calculation. References to some of the other methods are [25–28].

3.3   Evaporation

The canonical thermodynamical model or grand canonical model as described earlier calculates the properties of the collision averaged system that can be approximated by an equilibrium ensemble. Ideally, one would like to measure the properties of excited primary fragments after emission in order to extract information about the collisions and compare directly with the equilibrium predictions of the model. However, the time scale of a nuclear reaction (10^{-20}s) is much shorter than the time scale for particle detection (10^{-9}s). Before reaching the detectors, most fragments decay to stable isotopes in their ground states. Thus before any model simulations can be compared to experimental data, it is indispensable to have a model that simulates sequential decays. For the purposes of the sequential decay calculations, the excited primary fragments generated by the statistical model calculations are taken as the compound nucleus input to the evaporation code [29]. Hence, every primary fragment is decayed as a separate event.

We consider the deexcitation of a primary fragment of mass a, charge j and temperature T. The successive particle emission from the hot primary fragments is assumed to be the basic deexcitation mechanism. For each event of the primary breakup simulation, the entire chain of secondary breakup events is Monte Carlo simulated. The standard Weisskopf evaporation scheme is used to take into account evaporation of nucleons, d, t, He^3 and α. The decays of excited states via gamma rays were also taken into account for the sequential decay process and for the calculation of the final ground state yields. We have also considered fission as a deexcitation channel though for the nuclei of mass < 100 its role will be quite insignificant. The process of light particle emission from a compound nucleus is governed by the emission width Γ_ν at which a particle of type ν is emitted. According to Weisskopf's conventional evaporation theory [30], the partial decay width for emission of a light particle of type ν is given by

$$\Gamma_\nu = \frac{gm\sigma_0}{\pi^2 \hbar^2} \frac{(E^* - E_0 - V_\nu)}{a_R} \exp(2\sqrt{a_R(E^* - E_0 - V_\nu)} - 2\sqrt{a_P E^*})$$

(3.9)

Here m is the mass of the emitted particle, g is its spin degeneracy. E_0 is the particle separation energy which is calculated from the binding energies of the parent nucleus, daughter nucleus and the binding energy of the emitted particle and the liquid drop model is used to calculate the binding energies. The subscript ν refers to the emitted particle, P refers to the parent nuclei and R refers to the residual (daughter) nuclei. a_P and a_R are the level density parameters of the parent and residual nucleus respectively. The level density parameter is given by $a = a/16\,\mathrm{MeV}^{-1}$ and it connects the excitation energy E^* and temperature T through the following relations.

$$E^* = a_P T_P^2$$
$$(E^* - E_0 - V_\nu) = a_R T_R^2$$

(3.10)

where T_P and T_R are the temperatures of the emitting (parent) and the final (residual) nucleus respectively. V_ν is the Coulomb barrier which is zero for neutral particles and non-zero for charged particles. In order to calculate the Coulomb barrier for charged particles of mass $a \geq 2$ we use a touching sphere approximation,

$$V_\nu = \begin{cases} \dfrac{j_\nu(j_P - j_\nu)e^2}{r_i\{a_\nu^{1/3} + (a_P - a_\nu)^{1/3}\}} & \text{for } a_\nu \geq 2 \\[4mm] \dfrac{(j_P - 1)e^2}{r_i a_P^{1/3}} & \text{for protons} \end{cases}$$

(3.11)

where r_i is taken as 1.44 m. σ_0 is the geometrical cross-section (inverse cross-section) associated with the formation of the compound nucleus (parent) from the emitted particle and the daughter nucleus and is given by $\sigma_0 = \pi R^2$ where,

$$
R = \begin{cases} r_0\{(a_P - a_\nu)^{1/3} + a_\nu{}^{1/3}\} & \text{for } a_\nu \geq 2 \\ r_0(a_P - 1)^{1/3} & \text{for } a_\nu = 1 \end{cases}
\tag{3.12}
$$

with $r_0 = 1.2$ fm.

For the emission of giant dipole γ-quanta we take the formula given by [31]

$$
\Gamma_\gamma = \frac{3}{\rho_P(E^*)} \int_0^{E^*} d\varepsilon \rho_R(E^* - \varepsilon) f(\varepsilon)
\tag{3.13}
$$

where

$$
f(\varepsilon) = \frac{4}{3\pi} \frac{1 + \kappa}{m_n c^2} \frac{e^2}{\hbar c} \frac{(a_P - j_P)j_P}{a_P} \frac{\Gamma_G \varepsilon^4}{(\Gamma_G \varepsilon)^2 + (\varepsilon^2 - E_G^2)^2}
\tag{3.14}
$$

with $\kappa = 0.75$, and E_G and Γ_G are the position and width of the giant dipole resonance.

For the fission width we have used the simplified formula of Bohr–Wheeler given by

$$
\Gamma_f = \frac{T_P}{2\pi} \exp\left(-B_f/T_P\right)
\tag{3.15}
$$

where B_f is the fission barrier of the compound nucleus.

Once the emission widths are known, it is required to establish the emission algorithm which decides whether a particle is being emitted from the compound nucleus. This is done by first calculating the ratio $x = \tau/\tau_{tot}$ where $\tau_{tot} = \hbar/\Gamma_{tot}$, $\Gamma_{tot} = \sum_\nu \Gamma_\nu$ and $\nu = n, p, d, t, \text{He}^3, \alpha, \gamma$ or fission and then performing Monte Carlo sampling from a uniformly distributed set of random numbers. In the case that a particle is emitted, the type of the emitted particle is next decided by a Monte Carlo selection with the weights Γ_ν/Γ_{tot} (partial widths). The energy of the emitted particle is then obtained by another Monte Carlo sampling of its energy spectrum. The energy, mass and charge of the nucleus is adjusted after each emission. This procedure is followed for each of the primary fragment produced at a fixed temperature and then repeated over a large ensemble and the observables are calculated from the ensemble averages. The number and type of particles emitted and the final decay product in each event is registered and are taken into account properly keeping in mind the overall charge and baryon number conservation.

3.4 Discussion

Lastly it is to be noted that in computing partition functions no attention was paid to whether the particles are bosons or fermions and it was assumed that the classical Maxwell–Boltzmann limit applies. The qualitative argument is that the volumes used here are more than three times normal nuclear volume. At low temperature ($\sim 4\,\mathrm{MeV}$) where one might expect the approximation to fail, it survives because many composites appear, thus there is not enough of any particular species to make (anti)symmetrisation an important issue. At much higher temperature the number of protons and neutrons increase but as is well known the $n!$ correction takes the approximate partition function towards the proper one at high temperature. In a hypothetical world, the problem could get very difficult. Such a scenario would arise if the physics was such that at low temperature we only had neutrons and protons and no composites.

Chapter 4

Isoscaling

Isoscaling is an interesting and informative phenomenon seen in intermediate energy heavy ion collision [32–37]. This was observed in many cases. We will start with one case which was extensively studied both experimentally and theoretically. Isotope yields from central collisions of ^{112}Sn + ^{112}Sn, ^{112}Sn + ^{124}Sn, ^{124}Sn + ^{112}Sn and ^{124}Sn + ^{124}Sn were measured. Let us denote by R_{21} the ratio of yields of the same isotope (N, Z) from two different reactions 1 and 2, $R_{21}(N, Z) = Y_2(N, Z)/Y_1(N, Z)$. It is found that in many cases, the ratio is fitted extremely well by

$$R_{21}(N, Z) = C \exp(\alpha N + \beta Z) \qquad (4.1)$$

Several comments are necessary. Even in central collisions not all nucleons participate equally. Some leave the reaction zone very quickly as pre-equilibrium emission from periphery without or with rare collisions. Although reaction 1 was ^{112}Sn + ^{112}Sn after leaving out the pre-eqilibrium particles, participant numbers are taken to be $Z_1 = 75, N_1 = 93, A_1 = 168, N_1/Z_1 = 1.24$. Reaction 2 was ^{124}Sn + ^{124}Sn and number of participants are considered to be $Z_2 = 75, N_2 = 111, A_2 = 186, N_2/Z_2 = 1.48$. It is assumed that pre-equilibrium emission does not alter the N/Z ratio. Usual convention is that reaction 2 involves the more neutron rich pair. Equation (4.1) is particularly simple. For these two reactions 1 and 2 we see the following. For a fixed Z, $\log R_{2,1}(N, Z)$ is a linear function of N (in this example with a slope parameter $\alpha = 0.361$) and for fixed N it is a linear function of Z (with a slope parameter $\beta = -0.417$). This is

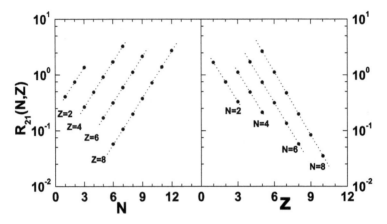

Fig. 4.1:   Ratios (R_{21}) of multiplicities of fragments (N, Z) where mass and charge of the fragmenting system for reaction 1 are $A_1 = 168$ and $Z_1 = 75$ respectively and these for reaction 2 are $A_2 = 186$ and $Z_2 = 75$. The left panel shows the ratios as function of neutron number N for fixed Z values, while the right panel displays the ratios as function of proton number Z for fixed neutron number N. The lines drawn through the points (circles) are best fits of the calculated ratios.

called isoscaling. The constant C is close to unity: $C = 1.163$. Plots of Eq. (4.1) are shown in Fig. 4.1.

The relationship of Eq. (4.1) readily follows from GCM. In previous chapters we established that in GCM a composite with N neutrons and Z protons has chemical potential $N\mu_n + Z\mu_p$ and the average number $\langle n_{N,Z} \rangle$ of the composite is $\langle n_{N,Z} \rangle = \exp(N\mu_n/T + Z\mu_p/T)\omega_{N,Z}$. Here μ_n is the neutron chemical potential and μ_p is the proton chemical potential; $\omega_{N,Z}$ is the one-particle partition function of the composite (N, Z). In reactions 1 and 2, the beam energy is the same so the temperature is expected to be the same. If we take the ratio of $\langle n_{N,Z} \rangle$ in the two reactions, the $\omega_{N,Z}$'s cancel out. The chemical potentials μ_n and μ_p for system 2 $(A_2 = 186, Z_2 = 75, N_2 = 111)$ will be different from the μ_n and μ_p for system 1 $(A_1 = 168, Z_1 = 75, N_1 = 93)$. The constant C should be nearly 1. Pre-equilibrium emission effects and sequential decay can cause deviation from 1. At creation, the composites have temperature T. This is the population generated by CTM or GCM. Often this already brings out the important physics. But this primary distribution can change by subsequent particle decay (evaporation) (Chapter 3). This may or may not significantly alter the primary distribution. It was found that at least in this particular example we are following, although the effect on primary distribution is quite noticeable the slope parameters α and β hardly change.

Let us call $\mu_n^{(2)}$ the neutron chemical potential and $\mu_p^{(2)}$ the proton chemical potential in the $N_2 = 111, Z_2 = 75$ system. Similarly, $\mu_n^{(1)}$ and $\mu_p^{(1)}$ be the the neutron and proton chemical potentials in the $N_1 = 93, Z_1 = 75$ system. Then $\alpha = (\mu_n^{(2)} - \mu_n^{(1)})/T \equiv (\delta\mu_n)/T$. Similarly $\beta = (\delta\mu_p)/T$. We will limit our discussion to α only by following the variation of $R_{2,1}$ along different N's but constant Z.

While in the present case GCM predicts isoscaling, CTM, which is a better model for finite nuclear systems, does not point to that readily. But CTM is easy to implement and one can numerically calculate $R_{2,1}$. Using CTM one finds that isoscaling is obeyed in this case and is in agreement with GCM. Further calculations reveal that isoscaling is obeyed if the composites are small compared to the system from which they come. In the example here N is restricted to 10 or less and Z to 8 or less while the emitting systems have either $N_1 = 93, Z_1 = 75$ or $N_2 = 111, Z_2 = 75$. In an experiment involving Ni on Be, reaction 1 was ^{58}Ni on ^{9}Be. Dissociating system was taken to be $N_1 = 35$ and $Z_1 = 32$. Reaction 2 was ^{64}Ni on ^{9}Be and the dissociating system was taken to be $N_2 = 41$ and $Z_2 = 32$. The beam energy was 140 MeV/n. It was found that $\log R_{2,1}(N, Z)$ for constant Z is linear with N for small Z. This is true up to $Z = 6$. For these low Z's the CTM and GCM results are also very close. At higher values of Z the plots begin to deviate from linearity, the slopes increase. The CTM and GCM curves (Figs. 4.3 and 4.2 respectively) begin to differ more and more and the experimental data are much better fitted with CTM. Fixing our attention to constant Z, in GCM the physics is dictated by $\alpha = \delta\mu_n/T$. CTM does not explicitly use a chemical potential, but we can deduce a μ by using the thermodynamic identity $\mu = (\partial F/\partial N)_{V,T}$ [2]. We can calculate the values of Q_{N_1,Z_1} and Q_{N_1-1,Z_1}. Since free energy F is just $-T \log Q$ we can compute $\mu_n^{(1)}$ from $-T(\log Q_{N_1,Z_1} - \log Q_{N_1-1,Z_1})$. Similarly we can compute $\mu_n^{(2)}$. Not surprisingly, the μ's and the $\delta\mu$ in CTM

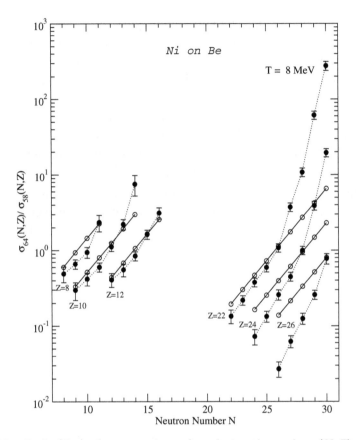

Fig. 4.2: Ratio (R_{21}) of cross-sections of producing the nucleus (N, Z) where reaction 1 is ^{58}Ni on ^{9}Be and reaction 2 is ^{64}Ni on ^{9}Be, both at 140 MeV/n beam energy. Experimental data with error bars are compared with theoretical results from grand canonical ensemble (hollow points). Dotted lines are drawn through experimental points and solid lines through calculated points.

and GCM agree remarkably well but in GCM the slope is solely dictated by $\delta\mu/T$ whereas this is not so in CTM in general. There have been efforts to deduce a value of the symmetry energy from an extracted value of $\delta\mu_n$. One formula that is used [32] is

$$\delta\mu_n \approx 4c_s \left[\left(\frac{Z_1}{A_1}\right)^2 - \left(\frac{Z_1}{A_2}\right)^2 \right] \qquad (4.2)$$

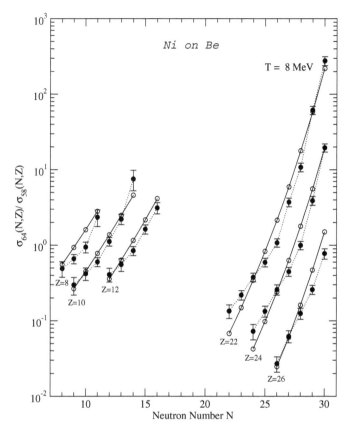

Fig. 4.3: The same as in Fig. 4.2 except that the calculation is done in the canonical ensemble. Agreement with data is far superior compared to that in Fig. 4.2.

Here c_s refers to the symmetry energy coefficient of Eq. (3.6). Derivation of the same quantity in the $T \to 0$ limit from CTM suggests a slightly different formula:

$$\delta\mu_n \approx 4c_s \left[\frac{Z_1}{A_1} - \frac{Z_1}{A_2} \right] \qquad (4.3)$$

However $\delta\mu_n$ as written above is temperature independent. In more rigorous derivations there will be a temperature dependence but it is quite complicated.

Chapter 5

A Model for Projectile Fragmentation

5.1 Introduction

In the introduction (Chapter 1) it was mentioned that in heavy ion collisions there are usually three different parts. One part is where nucleons from the projectile will meet on its way nucleons from the target. This part is the participant zone. There is also a projectile spectator. Nucleons in projectile spectator would not meet any nucleons from target if they followed straight line geometry. After the encounter, they move very nearly in the forward direction. The projectile spectator is also known as PLF (projectile like fragment). They have an excitation energy as the spectator has a crooked shape. Also some energy is transferred across the shearing plane. The excited PLF will expand to about three times the normal volume as it breaks up into pieces. The objective of this chapter is to calculate multifragmentation (multi means more than two) of the PLF. This has been the subject of intensive investigation. Similarly there is a target spectator which is not investigated here. In the lab they will be low energy (compared to PLF) fragments. Pictorially this is shown in Fig. 5.1.

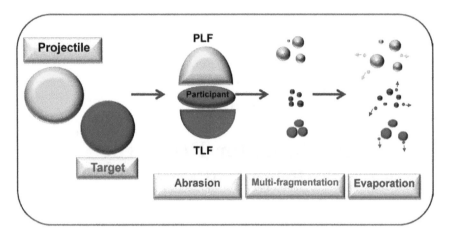

Fig. 5.1: Pictorial view of projectile spectator (PLF), target spectator (TLF) and participant fragmentation.

5.2 The model in a nutshell

As examples, we will consider Ni on Ta and Be at 140 MeV/nucleon [38], Xe on Al at 790 MeV/nucleon [39], ^{124}Sn and ^{107}Sn on ^{119}Sn [40] at beam energy 600 MeV/nucleon. Consider projectile with volume V_0, neutron number N_0 and proton number Z_0 hitting the target at impact parameter b. A certain fraction of the projectile is lost as participant. This can be calculated (Appendix A). What remains is the PLF, moving with a velocity close to the beam velocity. There is a probability of having N_s neutrons and Z_s protons in the PLF. This probability $P_{N_s,Z_s}(b)$ depends upon the impact parameter. We call this abrasion. The abrasion cross-section, where there are N_s neutrons and Z_s protons in the PLF, is labelled by σ_{a,N_s,Z_s}:

$$\sigma_{a,N_s,Z_s} = \int 2\pi b \, db \, P_{N_s,Z_s}(b) \qquad (5.1)$$

where suffix a denotes abrasion. This is stage 1 of the calculation [41].

An abraided system with N_s neutrons and Z_s protons has excitation. One characterises this by a temperature T. Trial values of temperature are tried to see what value seems the best. A more detailed scrutiny of T will be made in later sections of this chapter. The hot PLF will expand and break up into many composites (N, Z)

and nucleons. This break up is calculated using the CTM. The cross-section at this stage is called $\sigma^{pr}_{N,Z}$:

$$\sigma^{pr}_{N,Z} = \sum_{N_s,Z_s} n^{N_s,Z_s}_{N,Z} \sigma_{a,N_s,Z_s} \tag{5.2}$$

This finishes stage 2 of the calculation.

The composite (N, Z) after CTM is at temperature T. It can γ-decay to shed its energy but can also decay by light particle emission as described in Chapter 3. On the other hand, some higher mass nucleus can decay to this composite. Carrying through the "evaporation" finishes the calculation for comparing with experiments.

5.3   Some details

Consider the abrasion stage again. For an impact parameter b we can calculate the volume of the projectile that goes into the participant region. What remains in the PLF is V. If the original volume of the projectile is V_0, the original number of neutrons is N_0 and the original number of protons is Z_0, then the average number of neutrons in the PLF is $\langle N_s(b) \rangle = [V(b)/V_0]N_0$ and the average number of protons is $\langle Z_s(b) \rangle = [V(b)/V_0]Z_0$. These will usually be non-integers. Since in any event only integral numbers for neutrons and protons can materialise in the PLF we need to integerise the average numbers. If the average number is $\langle n \rangle$ we write $\langle n \rangle = n + \alpha$ where n is an integer and α is between 0 and 1. Then the probability of obtaining the value n is $(1 - \alpha)$ and the probability of obtaining the value $n + 1$ is α.

In Eq. (5.1) the upper limit of integration in b is $b_{max} = R_{proj} + R_{targ}$. If the projectile is larger than the target then $b_{min} = 0$. If the projectile is smaller than the target then $b_{min} = R_{targ} - R_{proj}$. At lower value of b there is no PLF. It is usual to evaluate the integral by a discrete sum. In Figs. 5.2 and 5.3 we compare some calculations with experimental data.

The next two figures present inclusive cross-sections: Fig. 5.4 for ^{58}Ni on ^{9}Be and Fig. 5.5 for ^{129}Xe on ^{27}Al. It is worthwhile remembering that in the Xe case the beam energy was much higher

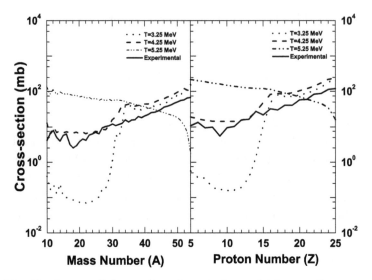

Fig. 5.2:   Total mass (left panel) and total charge (right panel) cross-section distribution for the ^{58}Ni on ^{9}Be reaction. The left panel shows the cross-sections as a function of the mass number, while the right panel displays the cross-sections as a function of the proton number. Theoretical calculations are shown for three temperatures to demostrate that the temperature 4.25 MeV is a good choice, others are not.

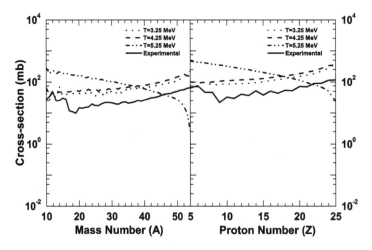

Fig. 5.3:   Same as Fig. 5.2 except that here the target is ^{181}Ta instead of ^{9}Be.

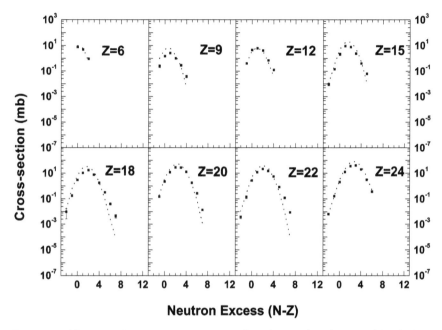

Fig. 5.4: Theoretical isotopic cross-section distribution (circles joined by dashed lines) for ^{58}Ni on ^{9}Be reaction compared with experimental data (squares with error bars). The temperature used for this calculation is 4.25 MeV.

(790 MeV/nucleon compared to 140 MeV/nucleon in the preceding cases).

What is remarkable is that although target projectile combinations are widely different and in one case shown here the beam energy is much higher than the other case the same temperature about 4.25 MeV seems to be appropriate. This is an example of limiting fragmentation. This is expected if binding energy per nucleon in each ion is much less than the beam energy/nucleon. Although the fits with data are pleasing, overall the fits for very peripheral collisions were not shown in the above figures. They are shown in Fig. 5.6 where near the upper limit of mass 58 or charge 28, the theoretical calculations underpredict cross-sections by large amounts. Near upper limit mass or upper limit charge can only come from extremely peripheral collision as it requires a very large PLF, almost

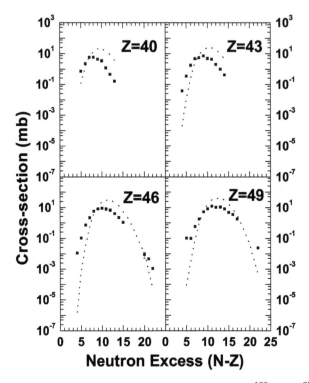

Fig. 5.5:   Same as Fig. 5.4 except that here the reaction is ^{129}Xe on ^{27}Al instead of ^{58}Ni on ^{9}Be. The temperature used for this calculation is 4.25 MeV.

the size of the projectile itself. With an assumed temperature of 4.25 MeV there will be too much of evaporative loss thus way underestimating high mass or charge cross-section. To correct this, we need to pay attention to "each" b, not just some "average" over the entire impact parameter range.

5.4   Impact parameter dependence of temperature

Experimental data on $\langle M_{IMF} \rangle$ as a function of Z_{bound} (Fig. 5.7) probably provide the strongest arguments for needing an impact parameter dependence of the temperature. Here $\langle M_{IMF} \rangle$ is the average multiplicity of intermediate mass fragments (z between 3 and 20) and Z_{bound} is the sum of all the charges coming from the PLF minus particles with $z = 1$. For ease of arguments we will

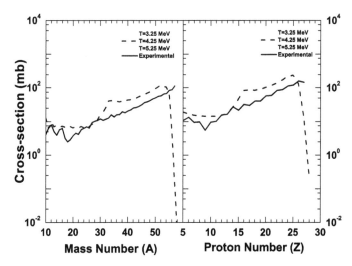

Fig. 5.6: Total mass (left panel) and total charge (right panel) cross-section distribution for the ^{58}Ni on ^{9}Be reaction including the regions coming from the very peripheral collisions. The left panel shows the cross-sections as function of mass number up to $A = 58$ (i.e. projectile mass), while the right panel displays cross-sections as function of proton number up to $Z = 28$ (i.e. proton number of projectile). The theoretical result at $T = 4.25$ MeV (dashed line) is compared with the experimental data (solid line). As stated in the text, very peripheral collision should have much lower temperature. The discrepancy between theory and experiment near the end is due to the fact that the same $T = 4.25$ MeV is used even for very peripheral collisions. The evaporative loss from the primary is far too great.

neglect in this section the difference between Z_{bound} and Z_s which is the sum of all the charges which originate from the PLF. A large value of Z_{bound} close to that of Z_0, the total charge of the projectile, signifies that the PLF is large and originated from peripheral collision (large b), whereas small Z_{bound} implies small b for nearly equal size ions. The following gross features of heavy ion collisions are well known. If the excitation energy of the dissociating system is low (equivalently low temperature), the system dissociates into one large piece and a small number of very light fragments. So $\langle M_{IMF} \rangle$ is small. As the temperature increases, very light as well as intermediate mass fragments appear at the expense of the heavy fragment. The $\langle M_{IMF} \rangle$ will grow as a function of temperature, will reach a peak and

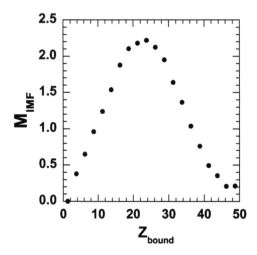

Fig. 5.7:   Mean multiplicity of intermediate mass fragments M_{IMF}, as a function of Z_{bound} obtained from ^{124}Sn on ^{119}Sn experiment at 600 MeV/nucleon performed by ALADIN collaboration in GSI.

then begin to fall since at high temperature only light particles are dominant. For a discussion of this see [42]. This is the rise and fall of $\langle M_{IMF} \rangle$ signifying the passage of the dissociating system from liquid phase to co-existence to vapour phase. For projectile fragmentation we are in the domain where $\langle M_{IMF} \rangle$ rises with temperature. For constant temperature $\langle M_{IMF} \rangle$ would continue to rise with Z_{bound}. Instead it reaches a peak and then begins to fall even though Z_{bound} increases. This can happen only if there is a drop in temperature as Z_{bound} hence b increases. The increase in size of the PLF is first compensated and then overcome by the drop in temperature. This is shown in Fig. 5.7 (courtesy of Prof. Trautmann).

5.5   Combine experiment and model to extract b dependence of temperature

The following iterative technique was used in [43] to deduce a temperature from experimental data of $\langle M_{IMF} \rangle$ vs Z_{bound}. Pick a b to get from abrasion a Z_s close to an experimental value of Z_{bound}. Guess a temperature T. A CTM calculation followed by evaporation is now done to get a Z_{bound} and $\langle M_{IMF} \rangle$. If the guessed value of T is too low, the calculated value of $\langle M_{IMF} \rangle$ will be too low for this Z_{bound} when confronted with data. In the next iteration the value of T will be raised. If on the other hand for the guessed value of T, the calculated value of $\langle M_{IMF} \rangle$ was too high, in the next iteration the value of T will have to be lowered. Of course when one changes T, the value of Z_{bound} also shifts but this change is smaller, and with a small number of iterations one can approximately reproduce an experimental pair: $Z_{bound}, \langle M_{IMF} \rangle$. Examples of this are given in Table 5.1 (for ^{124}Sn on ^{119}Sn) and 5.2 (for ^{107}Sn on ^{119}Sn). The first two columns are from experiment [40]; the numbers were given to us by Prof. Trautmann to whom we are grateful for assistance. The next two columns are the values of Z_{bound} and $\langle M_{IMF} \rangle$ we get from our iterations. These are accepted as close enough to the experimental pair. These are accepted as the value of b (sixth column) and the temperature T (fifth column). Table 5.2 provides a similar compilation for ^{107}Sn on ^{119}Sn.

Table 5.1:   Best fit and experimental values for ^{124}Sn on ^{119}Sn. The first two columns are data from experiment [40]. The next two columns are the values of Z_{bound} and M_{IMF} we get from our iterative procedure. These values are taken to be close enough to the experimental pair. These are obtained for a value of b (fifth column) and a temperature T (sixth column).

Experimental		Theoretical			
Z_{bound}	M_{IMF}	Z_{bound}	M_{IMF}	b (fm)	Required T (MeV)
11.0	1.421	11.080	1.424	2.912	6.398
15.0	1.825	15.094	1.818	3.625	6.108
20.0	2.145	19.984	2.131	4.4574	5.840
25.0	2.010	25.024	2.019	5.289	5.520
30.0	1.505	29.854	1.545	6.122	5.250
35.0	0.920	34.985	0.928	7.072	4.970
40.0	0.415	39.639	0.424	8.023	4.650
45.0	0.193	44.763	0.196	9.331	4.350
47.0	0.156	46.512	0.154	9.925	4.260
49.0	0.135	48.425	0.130	10.876	4.190

Table 5.2:   Same as Table 5.1, except that here the projectile is ^{107}Sn instead of ^{124}Sn.

Experimental		Theoretical			
Z_{bound}	M_{IMF}	Z_{bound}	M_{IMF}	b (fm)	Required T (MeV)
15.0	1.690	14.816	1.583	3.886	6.200
20.0	1.923	19.865	1.906	4.698	5.740
21.0	1.984	21.207	1.976	4.930	5.705
25.0	1.749	24.913	1.758	5.510	5.320
30.0	1.079	30.356	1.075	6.438	4.900
35.0	0.581	35.252	0.602	7.366	4.600
40.0	0.223	40.123	0.225	8.410	4.210
45.0	0.201	44.676	0.199	9.802	4.100
47.0	0.201	47.024	0.159	10.876	4.000

5.6   Towards a universal temperature profile

Knowing the temperature profile $T = T(b)$ in one case, say, ^{124}Sn on ^{119}Sn, can we predict what $T = T(b)$ will be in another case, say, ^{58}Ni on ^{9}Be? Both have $b_{min} = 0$ and $b_{max} = R_1 + R_2$ but we cannot expect the same functional form $T(b)$ in the two cases. In the first case, near $b = 0$, a small change in b causes a large fractional change in the mass of the PLF, whereas for Ni on Be a small change in b causes very little change in the mass of the PLF. Thus we might expect the temperature to change more rapidly in the first case near $b = 0$ whereas, in the second case the temperature may change very little since not much changed when b changed a little. This means that for Ni on Be, T would be slow to change in the beginning.

   We might argue that a measure of the wound that the projectile suffers in a heavy ion collision is

$$1.0 - A_s/A_0$$

(A_s is the mass number of the PLF and A_0 that of the projectile) and that the temperature depends upon the wound. Thus we explore an expansion: $T = D_0 + D_1[A_s(b)/A_0] + D_2[A_s(b)/A_0]^2 + \cdots$. We try fits to the "experimental" temperature profile given in Tables 5.1

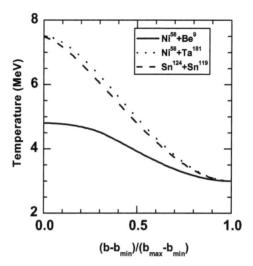

Fig. 5.8: Temperature profile for ^{58}Ni on ^{9}Be (solid line), ^{58}Ni on ^{181}Ta (dotted line) and ^{124}Sn on ^{119}Sn (dashed line) by considering $T = [7.5 - 4.5(A_s(b)/A_0)]$ MeV.

and 5.2. A linear fit appears to be adequate [43] with

$$T(b) = \left[7.5 - 4.5\left(\frac{A_s(b)}{A_0}\right)\right] \tag{5.3}$$

It is interesting to compare calculated temperature profiles for different pairs. In Fig. 5.8 we show this for ^{58}Ni on ^{9}Be, ^{58}Ni on ^{181}Ta and ^{124}Sn on ^{119}Sn. The formula does not depend upon beam energy or property of the target thus is compatible with limiting fragmentation.

5.7 Albergo formula for temperature

A prescription to extract a temperature from experimental data was proposed by Albergo [44]. This goes by the name Albergo formula. The prescription derives an expression for temperature from the ratio of yields in two pairs of nuclei resulting from the dissociation of the PLF (or participant). The yields are $Y(N_i, Z_i)$, $Y(N_i + \Delta N, Z_i + \Delta Z)$, $Y(N_j, Z_j)$, $Y(N_j + \Delta N, Z_j + \Delta Z)$. We seek to measure a ratio

$$R = \frac{R_{up}}{R_{down}} \tag{5.4}$$

where

$$R_{up} = Y(N_i, Z_i)/Y(N_i + \Delta N, Z_i + \Delta Z) \tag{5.5}$$

and

$$R_{down} = Y(N_j, Z_j)/Y(N_j + \Delta N, Z_j + \Delta Z) \tag{5.6}$$

From GCM the yield $Y(N_i, Z_i)$ of the ground state of the nucleus (N_i, Z_i) is given by

$$Y(N_i, Z_i) = \exp\left[\frac{1}{T}(N_i \mu_n + Z_i \mu_p)\right] gr_{N_i, Z_i} \tag{5.7}$$

Here $gr_{N_i, Z_i} = \frac{V}{h^3}(2\pi m_n T)^{3/2}(A_i)^{3/2}(2S(N_i, Z_i) + 1)\exp[BE(N_i, Z_i)/T]$ with $A_i = N_i + Z_i$ and $2S(N_i, Z_i) + 1$ the spin degeneracy factor and $BE(N_i, Z_i)$ the binding energy of the nucleus (N_i, Z_i).

We have corresponding equations for $Y(N_i + \Delta N, Z_i + \Delta Z)$, $Y(N_j, Z_j)$, and $Y(N_j + \Delta N, Z_j + \Delta Z)$. To get an expression for R, define

$$B = BE(N_i, Z_i) - BE(N_i + \Delta N, Z_i + \Delta Z) - BE(N_j, Z_j)$$
$$+ BE(N_j + \Delta N, Z_j + \Delta Z) \tag{5.8}$$

$$a = \frac{[2S(N_j, Z_j) + 1]/[2S(N_j + \Delta N, Z_j + \Delta Z) + 1]}{[2S(N_i, Z_i) + 1]/[2S(N_i + \Delta N, Z_i + \Delta Z) + 1]} \tag{5.9}$$

$$\eta = \left[\frac{A_j/(A_j + \Delta A)}{A_i/(A_i + \Delta A)}\right]^{3/2} \tag{5.10}$$

With straightforward algebra one obtains

$$T = \frac{B}{\ln(\eta a R)} \tag{5.11}$$

The factor η is of the order unity and in a different model the factor $3/2$ is replaced by 1. It is common to ignore η.

It has been pointed out that the Albergo formula [44] provides no correction for evaporation and empirical corrections have been

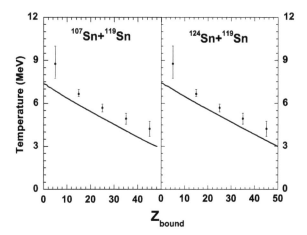

Fig. 5.9: Comparison of theoretically used temperature profile calculated by the formula $T(b) = 7.5 - 4.5(A_s(b)/A_P)$ (solid lines) with that deduced by Albergo formula from experimental data [40] (circles with error bars) for ^{107}Sn on ^{119}Sn (left panel) and ^{124}Sn on ^{119}Sn (right panel).

proposed [45]. Nonetheless, the universal temperature profile agrees quite well with Albergo temperatures. This is shown in Fig. 5.9.

5.8 Formulae for cross-sections

Discretize Eq. (5.1) with provision for variation of T with b. Divide the interval b_{min} to b_{max} into small segments of length Δb. Let the midpoint of the ith bin be $\langle b_i \rangle$ and the temperature at $\langle b_i \rangle$ be T_i. Then

$$\sigma_{a,N_s,Z_s} = \sum_i \sigma_{a,N_s,Z_s,T_i} \tag{5.12}$$

where

$$\sigma_{a,N_s,Z_s,T_i} = 2\pi \langle b_i \rangle \Delta b P_{N_s,Z_s}(\langle b_i \rangle) \tag{5.13}$$

PLF's with the same N_s, Z_s but different T_i's are treated independently. The rest of the calculation proceeds as before. If, after abrasion, we have a system (N_s, Z_s) at temperature T_i, CTM allows us to compute the average population of a composite with neutron number n and proton number z when this system breaks up (this composite is at a temperature T_i). Denote this by $M_{n,z}^{N_s,Z_s,T_i} \sigma_{a,N_s,Z_s,T_i}$. It then follows, summing over all the abraded (N_s, Z_s) that can yield (n, z), that the primary cross-section for (n, z) is

$$\sigma_{n,z}^{pr} = \sum_{N_s,Z_s,T_i} M_{n,z}^{N_s,Z_s,T_i} \sigma_{a,N_S,Z_s,T_i} \qquad (5.14)$$

Finally, evaporation from these composites (n, z) is considered (described in Chapter 4) before comparing with experimental data.

5.9 Some representative results

Some results, chosen arbitrarily, are shown here. They all use b dependent temperature: $T(b) = 7.5\,\mathrm{MeV} - [A_s(b)/A_0]\,\mathrm{MeV}$. Sometimes results do not change much if one uses instead a linearly b dependent temperature decreasing from $7.5\,\mathrm{MeV}$ at $b = 0$ to $3\,\mathrm{MeV}$ at b_{max}. In Fig. 5.10 experimental data [40] are compared with theoretical calculations.

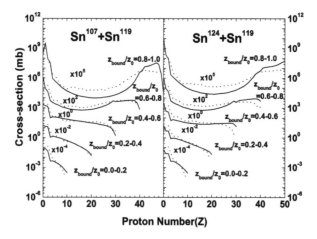

Fig. 5.10: Theoretical total charge cross-section distribution (solid lines) for ^{107}Sn on ^{119}Sn (left panel) and ^{124}Sn on ^{119}Sn reaction (right panel) sorted into five intervals of Z_{bound}/Z_0 ranging between 0.0 to 0.2, 0.2 to 0.4, 0.4 to 0.6, 0.6 to 0.8 and 0.8 to 1.0 with different multiplicative factors 10^{-4}, 10^{-2}, 10^0, 10^2, 10^5 respectively. The experimental data [40] are shown by dashed lines. The theoretical calculation is done using universal temperature profile given in Eq. (5.3).

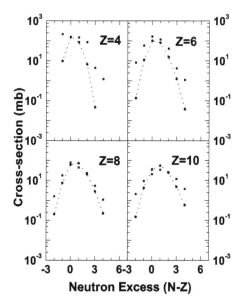

Fig. 5.11: Theoretical isotopic cross-section distribution (circles joined by dashed lines) for ^{107}Sn on ^{119}Sn reaction summed over $0.2 \leq Z_{bound}/Z_0 \leq 0.8$. The experimental data [40] are shown by black squires. The theoretical calculation is done using universal temperature profile given in Eq. (5.3).

In Fig. 5.11 the integrated isotopic distributions over the range $0.2 \leq Z_{bound}/Z_0 \leq 0.8$ for beryllium, carbon, oxygen and neon are plotted and compared with experimental result for ^{107}Sn on ^{119}Sn.

Next calculations are compared with data for ^{58}Ni on ^{9}Be and ^{181}Ta at beam energy 140 MeV/nucleon done at Michigan State University [46], see Figs. 5.12 to 5.15. Notice that data from peripheral collisions are well fitted.

Projectile fragmentation cross-sections of many neutron-rich isotopes have been measured experimentally from the ^{48}Ca and ^{64}Ni beams at 140 MeV per nucleon on ^{9}Be and ^{181}Ta targets [34]. A remarkable feature is the correlation between the measured fragment cross-section (σ) and the binding energy per nucleon (B/A). This observation has prompted attempts of parametrisation of cross-sections [47–49]. One very successful parametrisation is

$$\sigma = C \exp\left[\frac{B}{A}\frac{1}{\tau}\right] \tag{5.15}$$

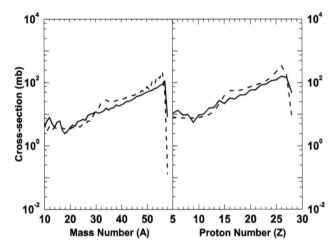

Fig. 5.12:   (a) Total mass and (b) total charge cross-section distribution for the ^{58}Ni on ^{9}Be reaction. The left panel shows the cross-sections as a function of the mass number, while the right panel displays the cross-sections as a function of the proton number. The theoretical calculation is done using temperature decreasing linearly with A_s/A_0 from 7.5 MeV to 3.0 MeV (dashed line) and compared with the experimental data [46] (solid line).

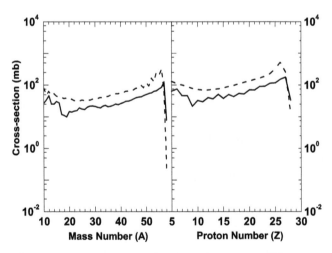

Fig. 5.13:   Same as Fig. 5.12 except that here the target is ^{181}Ta instead of ^{9}Be.

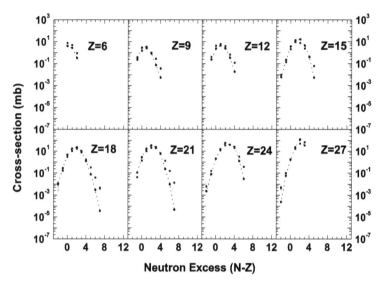

Fig. 5.14: Theoretical isotopic cross-section distribution (circles joined by dashed lines) for ^{58}Ni on ^{9}Be reaction compared with experimental data [46] (squares with error bars).

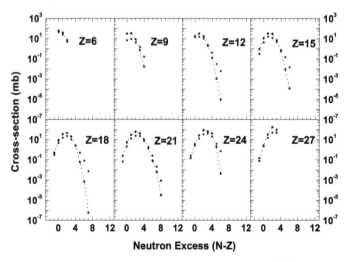

Fig. 5.15: Same as Fig. 5.14 except that here the target is ^{181}Ta instead of ^{9}Be.

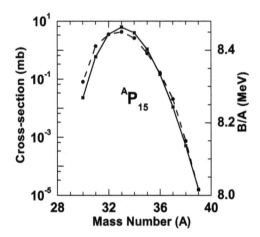

Fig. 5.16:   Fragment cross-section (circles joined by dotted line) for ^{64}Ni on ^{9}Be reaction and binding energy per nucleon (squares joined by solid line) plotted as a mass number for $Z = 15$ isotopes.

Here τ is a fitting parameter. In this parametrisation we have not considered the pairing energy contribution in nuclear binding energy. Here we have calculated production cross-sections of $Z = 15$ isotopes for ^{64}Ni on ^{9}Be reaction from projectile fragmentation model and plotted in log scale in Fig. 5.16 (circles joined by dotted line). The variation of the theoretical binding energy per nucleon for same isotopes of $Z = 15$ in linear scale is also shown in the same figure (squares joined by solid line). The similar trend of the cross-section curve (in log scale) and binding energy curve (in linear scale) confirms the validity of above parametrisation from our model [50]. By this method we can interpolate (or extrapolate) the cross-section of an isotope if the binding energy is known. We can also estimate the binding energy of an isotope by measuring its cross-section experimentally. This parametrisation also holds equally well for the reactions initiated by central collisions.

Average size of the largest cluster produced at different Z_{bound} values is calculated in the framework of projectile fragmentation model for ^{119}Sn and ^{124}Sn on ^{119}Sn reactions. In Fig. 5.17 the variation of Z_{max}/Z_0 (Z_{max} is the average number of proton content in the largest cluster) with Z_{bound}/Z_0 obtained from theoretical

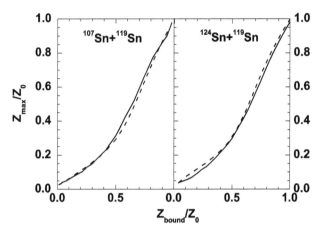

Fig. 5.17: Z_{max}/Z_0 as a function of Z_{bound}/Z_0 for ^{107}Sn on ^{119}Sn (left panel) and ^{124}Sn on ^{119}Sn (right panel) reaction obtained from projectile fragmentation model (dashed lines). The experimental results are shown by the solid lines.

calculation and experimental result are shown. Excellent agreement with experimental data is observed [51].

5.10 Discussion

The model of projectile fragmentation described above works satisfactorily though not extraordinarily well. It is based on a logically sensible prescription. One nice feature is that except for one parameter, the temperature, there is no free parameter. The parametrisation does not change from one pair of ions to another and remains unchanged over a wide range of beam energy. It might be possible to fine tune the temperature parameter and obtain some improvement.

The absolute values of cross-sections come out automatically. No normalisation is necessary. The estimate of the temperature used here can be obtained from a more basic theory [52].

The Copenhagen SMM model has been used to fit many data on projectile multifragmentation. The agreement is quite precise, but the approach is totally different. It starts with a reservoir of excited thermalised nuclear systems, labelled TNS. The most difficult part of the analysis is in selecting the TNS ensemble produced after the non-equilibrium stage of the reaction. Dynamical descriptions of the reaction starting from the entrance channel face the problem of making ambiguous predictions. Instead the practice is to start from the final state as represented by the experimental data. Search for a TNS distribution is made, that when used as input to SMM leads to the best possible reproduction of the data.

Chapter 6

Percolation and Lattice Gas Model

6.1 Percolation model

This is one of the earliest models used to interpret data on multi-fragmentation. The Purdue group observed a power law $\sigma(A_F) \propto A_F^{-\lambda}$ in the inclusive mass yield data for the reactions $p + \text{Kr}$ and $p + \text{Xe}$ at beam energies 80 to 300 GeV [53, 54]. The proton exits the nucleus after depositing some energy which leads to multifragmentation. The power law suggests the energy deposition takes the nucleus to the neighbourhood of critical point where it breaks up. Experimental data also exist for Au ions at beam energy 1 GeV/n on emulsions [55].

Let us see how we can use the well-known bond percolation model [56] to produce clusters. Consider a nucleus of N^3 nucleons which are in N^3 cubic boxes (one nucleon in each box). We do not distinguish between neutrons and protons. Proton charge is ignored. Nearest neighbours (they have a common wall) can bind together with a probability p_s. If p_s is 1 we have just one nucleus with N^3 nucleons and multiplicity M is 1. We can lower the value of p_s by pumping some energy into the system. Depending upon how the collision was, the amount of energy pumped in may be small, moderate or large. The sticking probability p_s will go down from the value 1 as more energy is pumped in. In the limit $p_s = 0$, we have only monomers; the multiplicity is now $M = N^3$. For a value of p_s between the two extremes 0 and 1 we can get clusters of different sizes

a ($a = 1$ is a monomer, $a = 2$ is a dimer, etc.) and the multiplicity of size a be denoted by n_a. The total multiplicity is $M = \sum_a n_a$ and the total number of nucleons is $N^3 = \sum_a a n_a$.

For an intermediate value of p_s we have to do Monte Carlo sampling to generate clusters. We need to generate many "events". Let us consider how to create one event. For a given value of p_s we need to sample every nearest neighbour bond to test whether it will be active (the two nucleons will be bound together) or inactive (the two nucleons are not directly bound together). For each nearest neighbour bond we call a random number x (value between 0 and 1) and compare it with the given value of p_s. If $p_s > x$ the bond is considered to be active. If $p_s < x$ the bond is considered to be inactive. Once we have gone through all the nearest neighbour bonds we check for clusters. If nucleon i is connected with nucleon j and j with k, then i, j and k are part of the same cluster even if there is no direct active bond between i and k. Thus in an event we can get the clusters a and their multiplicity n_a. Helpful tips for finding a and n_a once all the nearest neighbour bonds have been classified can be found in an appendix in [56].

To get a second "event" we repeat the same procedure with freshly generated random numbers for the nearest neighbour bonds. These n_a's will be different. A large number of events are accumulated. From these we can, for each a, get an average n_a, get an average total multiplicity M ($= \sum_a n_a$) and define a reduced multiplicity $n = M/A$ where $A = N^3$ is the number of nucleons. In each event we also pick the cluster with the largest value of a and compute the average value of the square of the largest a and call it a_{max}^2. The quantity below is called m_2:

$$m_2 = \frac{[\sum a^2 n_a - a_{max}^2]}{A} \qquad (6.1)$$

One can plot m_2 as a function of p_s but p_s is not a directly measurable quantity; alternatively one can plot m_2 as a function of n which is experimentally accessible. For $A = 216$ a theoretical curve of m_2 against n is shown in Fig. 6.1; a maximum in m_2 appears at about $n = 0.25$. Available data of Au on emulsion has the general shape

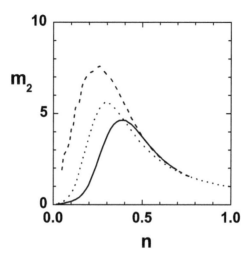

Fig. 6.1: Variation of m_2 with n with the lattice gas model at $D = 7$ (solid line) and $D = 8$ (dashed line) for a fragmenting system having $Z = 82$ and $N = 126$. The percolation model curve (dotted line) is for $A = 216$.

of Fig. 6.1. Theoretically one can check that the maximum becomes sharper and higher as one uses larger and larger particle numbers, like $7^3, 8^3, 9^3$, etc. This is regarded as a manifestation of a sharp second-order phase transition. According to the model one should expect that $n_a \propto a^{-\tau} f(\epsilon a^\sigma)$ where τ and σ are critical indices and ϵ is a measure of distance from the critical point. One has $f(0) = C$ so that at critical point, n_a should obey a power law.

The agreement between experimental m_2 and percolation model m_2 played a big role in the argument for continuous phase transition. But m_2 is not such a distinguishing feature. In Fig. 6.1 we have plotted the same m_2 but calculated with lattice gas model [57]. This also has similar behaviour and lattice gas model has a first-order phase transition.

There is no thermodynamic phase transition in percolation model. There is no temperature, no energy. The phase transition can be labelled percolative phase transition. At $p_s = 1$ ($n = 1/A \approx 0$) one can go from one side of the box to the opposite side through active bonds, i.e., stepping on nucleons which are part of the same cluster. This will continue for a while as p_s decreases till a point is reached where this will no longer be possible. We have reached the critical value p_c.

We will revisit percolation model later in the book.

6.2　Lattice gas model

Like in percolation model, we have $D^3 = N_T$ cubic boxes where we have to put in A nucleons but now $N_T \geq A$. A box can have one nucleon or none. Each cubic box has volume $\frac{1}{0.16}$ fm^3. The freeze-out density ρ is given by $\rho/\rho_0 = A/N_T$. The nuclear force is considered to be nearest neighbour interaction. In earlier applications of the model to investigate phase transition and cluster distributions, no distinction is made between neutrons and protons. The interaction between nearest neighbours is a constant value ϵ where ϵ is negative. At a temperature T one can try to obtain thermodynamic properties of this system like average energy, pressure, specific heat, phase transition, etc. This is the original lattice gas model. It is demonstrated in [1] that a grand canonical solution of this problem can be mapped onto the solution of the Ising model problem. Numerical work established that there is first-order transition in the model which leads to critical point at $T = 1.1275|\epsilon|$ and $\rho_c = 0.5\rho_0$.

We now introduce an extension of this model which is useful in nuclear physics [58]. We distinguish between neutrons and protons. The nearest neighbour neutron–proton interaction ϵ_{np} is set at -5.33 MeV. For proton–proton and neutron–neutron interaction, the usual values are $\epsilon_{pp} = \epsilon_{nn} = 0$. The last two values are chosen so that no bound n–n or p–p clusters emerge. If ϵ_{np} is set at -5.33 MeV then without Coulomb effect a large nucleus with $N = Z$ will have 16 MeV binding energy per nucleon. For this system, the lowest energy at $T = 0$ is obtained when sites are alternately populated by neutrons and protons so that all nearest bonds are neutron-proton type. Each nucleon (except the ones near the surface) has 6 nearest neighbour bonds, so neglecting surface effects the total energy is $-\frac{1}{2} \times 6 \times 5.33 \times A$ for a nucleus with $N = Z = A/2$. It is straightforward to include Coulomb interaction between protons. One simply computes $\sum_{i<j} \frac{e^2}{|\vec{r_i} - \vec{r_j}|}$ where i, j refer to protons.

There are N_T cubes in which we need to put N neutrons and Z protons and $N+Z = A$ is less than N_T. There are a very large number of possible configurations. A configuration is described by specifying which cubes are occupied by neutrons and which by protons. The empty ones are then automatically specified. A configuration is sometimes called an "event". Given the nearest neighbour interactions $\epsilon_{np}, \epsilon_{nn}, \epsilon_{pp}$ one can calculate the excitation energy E (with respect to the ground state). The occupation probability of this configuration is $\propto \exp(-E/T)$. A very large number of configurations which will reflect this property is generated by the Metropolis algorithm [59]. This is a very well-known method in statistical physics so here we merely state how, for our problem, we implement this. One can start with an *ad hoc* distribution of N neutrons and Z protons in N_T boxes but it would be less work if one starts with a reasonable distribution. Such a choice is prescribed in [58]. Other choices are possible. Starting from this configuration one tries to (1) switch the location of a neutron with that of a proton and (2) take a neutron or a proton from an occupied site to an empty site. If the energy goes down or stays constant by switching, the switch is accepted. If the energy goes up by ΔE then the switch is accepted with a probability $\exp(-\Delta E/T)$. A large number of switches (10^7) should be attempted before the first configuration is stored as an event. One can then start from the configuration of this event and start another set ($\approx 10^4$) of switches before accepting the next event. The large number of attempted switches before an event is accepted is to ensure that an event configuration is not biased by the previous event configuration from which it started. Once enough events are accumulated one can calculate average energy, specific heat and various other quantities some of which will be displayed.

To calculate clusters, further work is necessary. Kinetic energy will add $A \times \frac{3}{2}T$ to the total potential energy. We assign momenta to nucleons from Monte Carlo sampling of a Maxwell–Boltzmann distribution at temperature T. Two nearest neighbours will be bound to each other if their relative kinetic energy is less than the attraction between the two, i.e. if $\vec{p_r}^2/2\mu + \epsilon < 0$. Here $\vec{p_r} = \frac{1}{2}(\vec{p_1} - \vec{p_2})$, and $\mu = m/2$. Since we take $\epsilon_{nn} = \epsilon_{pp} = 0$ two nearest neighbour neutrons or two nearest neighbour protons cannot directly bind but nearest neighbour neutron–proton can ($\epsilon_{np} = -5.33$ MeV). We apply the same rule as in percolation. If i is directly bound to j and j is directly bound to k, then i, j and k are considered to be part of the same cluster even if there is no direct bond between i and k. This is discussed in greater detail in [60] and [61]. The prescription given here is enough to compute clusters but we will pursue the bonding formula a bit more. Figure 6.1 shows some results obtained from lattice gas model.

Let us return to the bonding prescription. It can be cast in a form where it can be compared to a well-known formula. The prescription for bonding is $\vec{p_r}^2/2\mu + \epsilon < 0$. Using the fact that in a Maxwell–Boltzmann distribution the relative motion is also Maxwell–Boltzmann, the bonding equation is given by

$$p = 1 - \frac{4\pi}{(2\pi\mu T)^{3/2}} \int_{\sqrt{2\mu|\epsilon|}}^{\infty} \exp(-p_r^2/2\mu T)p_r^2 dp_r$$

$$= 1 - \frac{2}{\sqrt{\pi}} \int_{|\epsilon|/T}^{\infty} e^{-q}q^{1/2}dq \tag{6.2}$$

In Coniglio–Klein work the equation for bonding between nearest neighbours [56, 62] is

$$p = 1 - \exp(-|\epsilon|/2T) \tag{6.3}$$

The two prescriptions are compared in Fig. 6.2. They are very close. In Coniglio–Klein model, the thermodynamic critical point ($T_c = 1.1275|\epsilon|, \rho_c = 0.5\rho_0$) is also the percolative critical point. We should expect the same here. However percolation sets in not just at the thermodynamic critical point, rather along a continuous line in the ρ–T plane from ρ_c to ρ_0. This is shown in Fig. 6.3.

Many more aspects of phase transition in lattice gas model can be found in [63]. Very likely at intermediate energy we never get close to the critical point. All experimental results point to a dissociation density less than $0.5\rho_0$ so at dissociation the system at most goes through a first-order phase transition [64]. One convenient aspect of the lattice gas model is that, after the composite calculation there is no evaporation to consider. Composites are particle stable. This is a

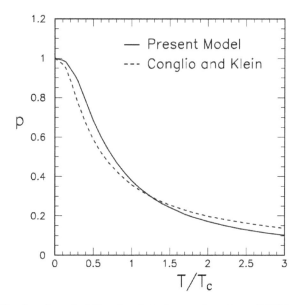

Fig. 6.2: The bond probability is plotted as a function of T/T_c where $T_c = 1.1275|\epsilon|$. The solid curve is from the model described here (Eq. (6.2)). The dashed curve is the Coniglio–Klein formula (Eq. (6.3)).

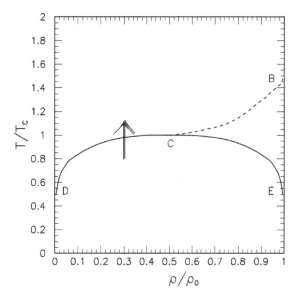

Fig. 6.3: Phase diagram of three-dimensional lattice gas model. The line DCE is the coexistence curve. CB is the percolation line. The arrow demonstrates the crossing of the coexistence line as the beam energy for central collisions increases.

non-negligible advantage. Not many data fittings with the lattice gas model are available; some applications can be found in [65]. Mean-field calculations can be done for the lattice gas model [66]. The attractive feature of the mean-field calculation is that it is easy, gives phase transition and can give an approximate location of the critical point.

We will have occasion to use the lattice gas model again.

Chapter 7

Bimodality

Of the many experimental data that have been proposed as signatures of phase transition and its order, bimodality is a late entrant in intermediate energy heavy ion collision. We mention here three papers [67–69] which provide theoretical exposition and some experimental background. Let us refer to CTM for the system of one kind of particles (Chapter 2). We can have monomers ($k = 1$), dimers ($k = 2$), trimers ($k = 3$) and so on. We will consider (1) a system of 200 particles (k going from 1 to 200) and also a larger system of 2000 particles (k going from 1 to 2000). In events there are particles of many sizes. Let us call $P_{max}(k)$ the probability that k is obtained as the highest mass in an event. This is given by Eq. (2.9). The average value of the largest mass is given by $\langle k_{max} \rangle = \sum P_{max}(k) \times k$. The intensive quantity is $\langle k_{max} \rangle /A$ where A is the total number of particles. This is plotted in Fig. 2.2 as a function of temperature for $A = 200$ and $A = 2000$.

For the 200-particle system, at low temperature most of the particles are in a large blob (this blob is considered to be liquid). $\langle k_{max} \rangle /A$ is close to 1. At higher temperature the same quantity is close to 0 and we are in gaseous phase. When we go to the 2000-particle system the picture becomes much sharper. In the limit $A \rightarrow \infty$ the order parameter is 1 in the liquid phase and goes through a sudden drop to 0 as it crosses the transition temperature ("boiling temperature").

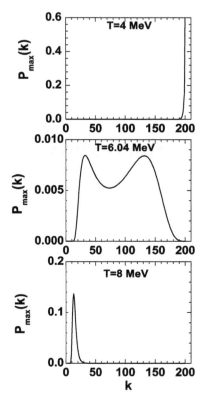

Fig. 7.1:   Probability that the largest cluster has k nucleons plotted as a function of k for a fragmenting system of mass 200 at constant freeze-out volume $6V_0$ but three different temperatures 4 MeV (upper panel), 6.04 MeV (middle panel) and 8 MeV (lower panel).

Bimodality is a result of the singularity (infinite system) being replaced by the smearing (finite system). How does it show up? For the system of 200 particles we plot $P_{max}(k)$ at $T = 4$ MeV and it is limited to high k side (liquid); see Fig. 7.1. At $T = 8$ MeV it is limited to low k side (gas); but at an intermediate value of temperature, in a small temperature range there exists a double hump. The temperature where the two humps have equal heights is a transition temperature. Although this has been demonstrated here with CTM only, in [69] this is demonstrated with the lattice gas model.

It is usual to identify the temperature at which c_v maximises as the phase transition temperature. We might call the temperature

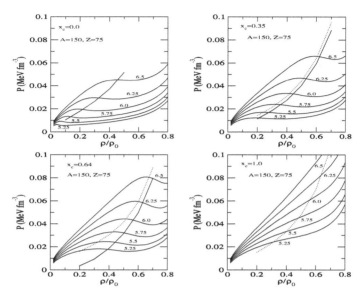

Fig. 7.2: EoS (p–ρ) for hypothetical nuclei with a varying strength of the Coulomb force x_c from zero (upper left) to the physical value $x_c = 1$ (lower right). The temperatures (in MeV) of the isothermals are indicated. Along the solid line the two peaks in $p_1(z_1)$ as a function of z_1 are of the same height; $p_1(z_1)$ is the probability that the maximum charge is z_1. The dotted line denotes the line where the specific heat c_v is maximum. The two lines coincide for $x_c = 0$.

where the two humps are of equal height the bimodal temperature. In Fig. 7.2 we draw EoS for a fictitious system with $A = 150$, $Z = 75$ (now we are considering two kinds of particles, Chapter 3). Our objective is to investigate how the Coulomb force affects the EoS. To study this, we use, following [70], a variable strength factor x_c. If $x_c = 0.0$ proton charge is set at 0. If $x_c = 1$ protons have their full charge. We also use $x_c = 0.35$ and 0.64 to see how properties change with the strength of the Coulomb force. For $x_c = 0$, the maximum of c_v and the bimodal temperature fall on top of each other. At $x_c = 0.35$ the bimodal temperatures and c_v maxima are still very close but not identical. The differences are substantial at $x_c = 0.64$. At $x_c = 1.0$, bimodality has disappeared but c_v maxima are still there.

Bimodal behavior studied from BUU transport model will be described in Chapter 15.

Chapter 8

Phase Transition

8.1 Phase transitions at intermediate energies

We have already talked about phase transitions in heavy ion collisions at intermediate energies; see Chapters 2, 6 and 7. The overall impression from Chapter 2 (although it was a simple model) was that one expects, in the nuclear case, a first-order phase transition. In the first part of Chapter 6, we introduced a power law i.e., $n_a \propto a^{-\lambda}$. Here a is the mass number of a fragment resulting from multifragmentation and n_a is the multiplicity of a. Percolation model has a second-order continuous phase transition where the average yield n_a of mass number a is expected to go like $n_a \propto a^{-\tau} f(\epsilon a^{\sigma})$. Here τ and σ are critical indices and ϵ is a measure of distance from the critical point. At the critical point $\epsilon = 0$ and f is independent of a and a power law would follow. In many models (CTM, GCM and LGM), temperature is a parameter, the critical point has a well defined temperature and the formula that is used is called the Fisher formula [71]

$$n_a = a^{-\tau} f(a^{\sigma}(T - T_c)) \tag{8.1}$$

$f(a^{\sigma}(T-T_c))$ is called the scaling law. The form of f is not postulated but that it depends on the combination $a^{\sigma}(T - T_c)$.

Towards the end of the nineties, EOS collaboration made detailed studies of break up of Au nuclei at 1 GeV/n on carbon target [72–76]. In a series of papers they not only give details of the experiment but also explain how they obtain the values of critical exponents. Instead of $T - T_c$ they use $m - m_c$ as the distance from the critical point where m is the total multiplicity and m_c is the critical multiplicity. From experimental data they deduce $m_c \approx 26$, $\tau \approx 2.2$ and σ between 0.63 to 0.70. Scaling is approximately obeyed but one would have liked tighter scatter for different a's. Incidentally, percolation gives a value of σ to be 0.45 [73].

One might argue that the Fisher model is the correct model for multifragmentation but the finiteness of the fragments and the total fragmenting system bring in non-ideal fits. The fragments a should be large but should also be only a small fraction of the fragmenting system. That situation is not met in the laboratory. But other points have been raised against evidence of criticality. The power law is no longer taken as a proof of criticality. There are many systems which exhibit a power law: mass distributions of asteroids in the solar system, debris from the crushing of basalt pellets [77] and the fragmentation of frozen potatoes [78]. The most compelling evidence of first-order phase transition came from experiments at GSI from Au on Au collisions at 600 MeV/n [79]. Experimentally they obtained a caloric curve which looks exactly like a caloric curve for first-order liquid–gas phase transition. This we quote from the conclusion of the paper: "From the observed fragment and neutron distribution the masses and excitation energies of the decaying prefragments were determined. A temperature scale was derived from observed yield ratios of He and Li isotopes. Rising first strongly with increasing excitation energy, the isotope temperature stays rather constant at a value of about 5 MeV for excitation energies between 3 and 10 MeV per nucleon. For higher excitation energies again an increasing temperature is found." Another interesting conclusion can be drawn from this work. "Depending on the low density equation of state the freeze-out in this vapour regime may be charactrerized by a density between 0.15 and 0.3 of normal nuclear density."

It is logical to ask the question: if there is only first-order phase transition in a model, then how badly will Eq. (8.1) be violated?

This was tested with CTM [80]. For an imagined nucleus, CTM calculations were done. Particle multipicity n_a's were calculated. Considering these as the "data", "best" estimates for T_c, τ and σ were obtained. It was then checked if, with these derived constants, $n_a a^\tau$ when plotted against $a^\sigma (T - T_c)$ collapse on the same curve for all a's. They of course do not but the smearing while not negligible, is not large and could have been attributed to finite size correction. Thus one might also advance these results as arguments for continuous phase transition. These graphs can be seen in [80] and p. 12 of [13]. Incidentally, the derived T_c does turn out to be close to the first-order phase transiton temperature of the CTM calculation.

Given n_a's to deduce "best" values of τ, T_c, σ is not straightforward but prescriptions can be found in [80, 81].

8.2 A signature that can establish the presence or absence of first-order phase transition

At intermediate energies there are basically two problems with trying to establish the order of phase transition. One is that the dissociating system is not large; the other is that there is Coulomb interaction which, in addition to limiting the size of nuclei, tends to smear out expected effects that appear at phase transition. We showed in Chapter 7 that in first-order phase transition, the effect of finite size is to produce a bimodal distribution. Without Coulomb interaction, the bimodal temperature (the temperature at which the two humps of the bimodal distribution have equal heights) and the temperature at which C_v maximises coincide. The appearance of a maximum in C_v marks the boiling point of this finite system and is a hallmark of first-order phase transition. The presence of Coulomb interaction complicates the situation. In Chapter 7 this was studied using a strength factor x_c. If $x_c = 0$, protons have no charge. If x_c is 1, protons carry their full charge. Bimodality and C_v are studied progressively as x_c is increased from 0 to 1. As mentioned, for $x_c = 0$, bimodality appears and the specific heat C_v hits a maximum value at the bimodal point. For small values of x_c, bimodality region shrinks and the maximum value of C_v is close to the bimodal temperature but not identical. Bimodality disappears before reaching $x_c = 1$ but the usual behaviour of C_v reaching a maximum continues till $x_c = 1$. So if we could measure C_v we would see vestiges of first-order phase transition even with the usual Coulomb force. But since measuring C_v is not a practical suggestion, is there some other measurable quantity that also maximises when C_v does?

The answer turns out to be rather simple. Theoretical modelling with CTM predicts that the derivative of total multiplicity with respect to temperature displays a maximum which coincides with the maximum of C_v. Since temperature increases with increasing beam energy, the maximum can be located in experiments. We will give some examples with CTM before moving to other models.

8.3   Illustrations with CTM

In this section we follow closely [82]. We use $n_{i,j}$ as the multiplicity of the composite with neutron number i and proton number j. Total multiplicity is then $M = \sum_{i,j} n_{i,j}$. For a dissociating system with $Z = 82$ and $N = 126$ (representing a moderately heavy system) we compute fragmentation at different temperatures, compute all $n_{i,j}$, sum them up and plot M as a function of T. This is shown in the left upper corner of Fig. 8.1. The calculation is done using T of course; the values of E^*/A are obtained from theory for this T. In experiments the Albergo formula could be used to estimate the temperature. From M against T we compute numerically dM/dT (top right corner). The maximum of dM/dT is approximately at 5 MeV which is seen to be the same temperature where C_v is maximum (see Fig. 8.3). Hence the phase transition temperature is approximately 5 MeV, the same as reported in [79]. The lower 2 diagrams in Fig. 8.1 use one kind of particles (Chapter 2) to demonstrate how Coulomb effects smear out peaks. Figure 8.2 shows results for a much lighter system.

In Fig. 8.3 we plot dM/dT and C_v on the left side for $Z = 82$, $N = 126$. On the right side the same quantities for $A = 208$ (one kind of particles and no Coulomb) just to show once again that peaks are much sharper without Coulomb. Figure 8.4 is similar to Fig. 8.3 except that it is for a smaller system. Figure 8.5 shows that entropy increases more rapidly near the maximum of dM/dT

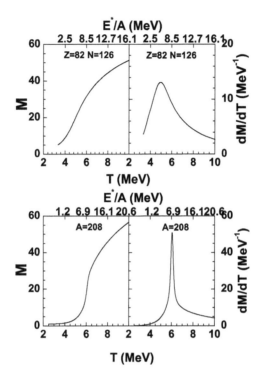

Fig. 8.1: Variation of multiplicity M (left panels) and dM/dT (right panels) with temperature (bottom x-axes) and excitation per nucleon (top x-axes) from CTM calculation for fragmenting systems having $Z = 82$ and $N = 126$ (top panels). Bottom panels represent the same but for hypothetical system of one kind of particles with no Coulomb interaction but the same mass number ($A = 208$). E^* is $E - E_0$ where E_0 is the ground state energy of the dissociating system in the liquid drop model.

than elsewhere. It is well known that composites from CTM are excited and will undergo sequential two-body decay which will change the total multiplicity. This will not change our conclusion. In fact sequential decay makes the peak in dM/dT sharper. This is shown in Fig. 8.6.

In many experiments, M_{IMF} is measured instead of total M. In the case studied here, the maximum of dM_{IMF}/dT does not accurately coincide with the maximum of C_v most likely because all composites are included in computing C_v but not in M_{IMF}; this is shown in Fig. 8.7.

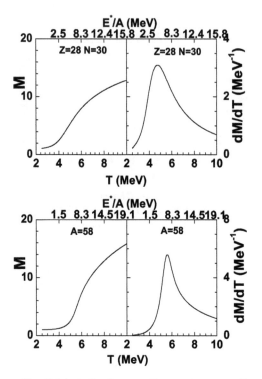

Fig. 8.2: Same as Fig. 8.1 but the fragmenting systems are $Z = 28$ and $N = 30$ (top panels) and $A = 58$ (bottom panels).

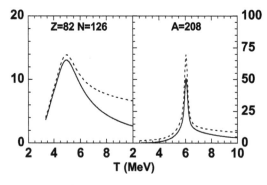

Fig. 8.3: Variation of dM/dT (solid lines) and C_v (dashed lines) with temperature from CTM for fragmenting systems having $Z = 82$ and $N = 126$ (left panel) and for hypothetical system of one kind of particles with no Coulomb interaction of mass number $A = 208$. To draw dM/dT and C_v in the same scale, C_v is normalised by a factor of $1/50$.

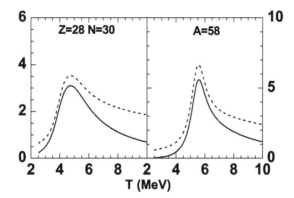

Fig. 8.4: Same as Fig. 8.3 but the fragmenting systems are $Z = 28$ and $N = 30$ (left panel) and $A = 58$ (right panel).

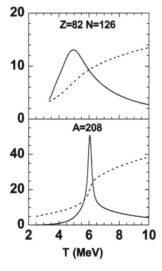

Fig. 8.5: Variation of entropy (dashed lines) and dM/dT (solid lines) with temperature from CTM for fragmenting systems having $Z = 82$ and $N = 126$ (top panel) and for hypothetical system of one kind of particles with no Coulomb interaction of mass number $A = 208$ (bottom panel). To draw S and dM/dT in the same scale, S is normalised by a factor of $1/20$ for $Z = 82$ and $N = 126$ system and $1/50$ for hypothetical system of one kind of particles.

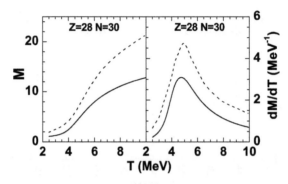

Fig. 8.6: Effect of secondary decay on M (left panel) and dM/dT (right panel) for fragmenting systems having $Z = 28$ and $N = 30$. Solid lines show the results after the multifragmentation stage (calculated from CTM) whereas dashed lines represent the results after secondary decay of the excited fragments.

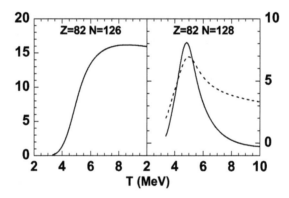

Fig. 8.7: Variation of intermediate mass fragment (IMF) multiplicity M_{IMF} (left panels) and first-order derivative of IMF multiplicity dM_{IMF}/dT (right panels) with temperature from CTM calculation for fragmenting systems having $Z = 82$ and $N = 126$. Variation of C_v with temperature (T) is shown by dashed line in right panel. To draw dM_{IMF}/dT and C_v in the same scale, C_v is normalised by a factor of $1/100$.

8.4 Multiplicity derivative in other models

Using CTM we have shown that the derivative of M going through a maximum implies that the system goes through a first-order phase transition. We want to check that in another model which also has a first-order phase transition. We also like to check that in a model that

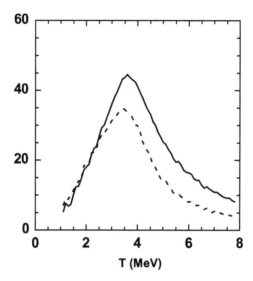

Fig. 8.8: Variation of dM/dT (solid lines) and C_v (dashed lines) with temperature from lattice gas model at $D = 8$ (see text) for fragmenting system having $Z = 82$ and $N = 126$. To draw dM/dT and C_v in the same scale, C_v is normalised by a factor of $1/10$; dM/dT has unit of MeV^{-1}.

does not have a first-order phase transition no maximum appears in M-derivative [57]. For this we use the percolation model, and for the other model which has a first-order phase transition we choose the lattice gas model. Let us now turn to percolation model. Both have been discussed in Chapter 6 and we will use the knowledge gained in there to our advantage. We first deal with the lattice gas model then percolation model. In Fig. 8.8 the appearance of a maximum in C_v and also in dM/dT demonstrate the existence of a first-order phase transition in the lattice gas model. For percolation we use 216 nucleons which are put in 6^3 cubical boxes. There is only one parameter in percolation model. This is the probability p_s that two nearest neighbours (which have a common wall) will bind together. If p_s is 1 we have one nucleus with 216 nucleons and M is 1. If p_s has the other extreme value 0, there will be 216 monomers and M is 216. For values of p_s between these two extremes, M is calculated by Monte Carlo sampling. We find it more convenient to use $p_b = 1 - p_s$

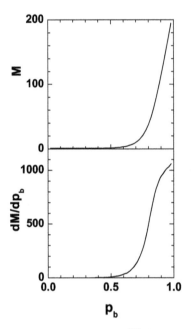

Fig. 8.9:   Variation of M (upper panel) and $\frac{dM}{dp_b}$ (lower panel) with p_b obtained from bond percolation model for a system of 6^3 nucleons.

and use p_b as the variable. If p_b is 0, then $M = 1$. If $p_b = 1$, M achieves the maximum value.

A plot of M and dM/dp_b is shown in Fig. 8.9; dM/dp_b has no maximum. Contrast this with the top part of Fig. 8.1 (CTM) and Fig. 8.8 (lattice gas model). Thus we come to the conclusion that the appearance of maximum in M-derivative is a signature that the dissociating system undergoes a first-order phase transition; the absence of the maximum rules out a first-order phase transition.

Very recently, an alternative signature for first-order phase transition was proposed [83]. This requires investigating the derivative $d\langle A_{max}\rangle/dT$ or $d\langle Z_{max}\rangle/dT$ where $\langle A_{max}\rangle$ is the largest mass number after dissociation and $\langle Z_{max}\rangle$ is the largest charge number after dissociation. The appearance of a maximum would signal a first-order phase transition. Initial results with a simplistic model are encouraging, but it needs to be backed by a more realistic model of fragmentation.

8.5 Experimental verification of multiplicity derivative

Very recently peak in multiplicity derivative is experimentally verified as a signature of nuclear liquid–gas phase transition [84]. The experiment is performed in TAMU $K = 500$ superconducting cyclotron by measuring the quasiprojectile reconstructed from the reaction of ^{40}Ar on ^{27}Al, ^{48}Ti and ^{58}Ni at $47\,\text{MeV/nucleon}$. In each case the peak and width of the dM/dT distribution are similar to that of specific heat.

Chapter 9

Incorporating Fermionic and Bosonic Statistics

The objective of this chapter is to quantitatively estimate errors incurred in using low density and/or high temperature limit to calculate average multiplicity of composites in the canonical model as used here. We did not use fermionic or Bose–Einstein statistics for nucleons or composites. If the partition function of 1 particle of the species i is ω_i (which is assumed to be known accurately) then it was assumed that the partition function of n_i such particles is given by $\frac{\omega_i^{n_i}}{n_i!}$. It did not matter if the species i is a fermion or a boson, $n_i!$ is the proper correction to apply. This is of course quite standard. In a case typical in intermediate energy heavy ion collision, we will treat fermions as fermions, bosons as bosons, calculate the n_i's and compare with values obtained with the standard prescription as used here so far. The results are presented in Sec. 9.2. The reader can directly go to that section. For completeness, we explain how we get the numbers in a calculation maintaining the fermionic (bosonic) character of the species.

9.1 Methodology

In Chapters 2 and 3 we introduced the canonical thermodynamic models. In Chapter 2 starting from Eq. (2.1), i.e.,

$$Q_A = \sum \prod \frac{(\omega_k)^{n_k}}{n_k!} \tag{9.1}$$

we arrived at a recursion formula, Eq. (2.4),

$$Q_A = \frac{1}{A} \sum_{k=1}^{A} k\omega_k Q_{A-k} \qquad (9.2)$$

Here k is the mass number of the composite ($k =$ number of nucleons in the composite), ω_k is the partition function of 1 such composite and n_k is the number of particles of type k. In Eq. (9.1) $A = \sum kn_k$.

Jennings pointed out that the grand partition function is a generator for the canonical partition function and can give a recursion formula like Eq. (9.2) [85, 86]. The grand partition function is defined as

$$Z_{gr}(\beta, \tilde{z}) = \sum_{A=0}^{\infty} \tilde{z}^A Q_A(\beta) \qquad (9.3)$$

where $Q_A(\beta)$ is the canonical partition function and $\tilde{z} = e^{\beta\mu}$ is the fugacity. From Eq. (9.3) we see that the grand canonical function is the generating function for the canonical partition function. Expand the log of the grand canonical function in power of the fugacity,

$$\ln Z_{gr}(\beta, \tilde{z}) = \sum_{k=1}^{\infty} \tilde{z}^k x_k \qquad (9.4)$$

Combining these two equations we have

$$\sum_{A=0}^{\infty} \tilde{z}^A Q_A(\beta) = \exp\left[\sum_{k=1}^{\infty} \tilde{z}^k x_k \right] \qquad (9.5)$$

The canonical partition function can be obtained by expanding the right hand side of Eq. (9.5) and equating powers of $\tilde{z}$. Thus

$$Q_A(\beta) = \sum_{n_k} \prod_k \frac{x_k^{n_k}}{n_k!} \qquad (9.6)$$

where the sum is over all partitions n_k of A. This looks like just Eqs. (9.1) and (2.1). There we had ω_k which was just the 1-particle partition function of the cluster with mass number k. Here we did not have to specify what x_k is. The easiest way to arrive at a recursion relation like Eq. (9.2) is to differentiate Eq. (9.5). We get

$$\sum_{A=0}^{\infty} A\tilde{z}^{A-1}Q_A(\beta) = (k\tilde{z}^{k-1}x_k)\exp\left[\sum_{k=1}^{\infty}\tilde{z}^k x_k\right]$$

$$= \left(\sum_{k=1}^{\infty}k\tilde{z}^{k-1}x_k\right)\left[\sum_{A=0}^{\infty}\tilde{z}^A Q_A(\beta)\right] \quad (9.7)$$

Equating powers of $\tilde{z}$ we obtain in a simple way the recursion for canonical partition functions

$$AQ_A(\beta) = \sum_{k=1}^{A}kx_k Q_{A-k}(\beta) \quad (9.8)$$

We write the above equation in two different though obvious ways. The first one is to build Q_A starting from $Q_0 = 1$ to a large value Q_A. This is

$$Q_A(\beta) = \frac{1}{A}\sum_{k=1}^{A}kx_k Q_{A-k}(\beta) \quad (9.9)$$

The other one is

$$A = x_1\frac{Q_{A-1}}{Q_A} + 2x_2\frac{Q_{A-2}}{Q_A} + 3x_3\frac{Q_{A-3}}{Q_A} + \cdots + Ax_A\frac{Q_0}{Q_A} \quad (9.10)$$

A formula like this is useful in the more complicated case that we will be treating. Next problem is to know the values of x_k. To know x_k we need to know the log of the grand partition function for the problem to be tackled. We will give some examples. Although the grand canonical distribution is used in a formal manner to obtain x_k, they can be used to obtain the canonical partition function and their use does not mean that the particle number is not fixed.

We first consider only protons filling up orbitals $i, j, k, \ldots$ in a box. Now

$$\ln Z_{gr}(\beta\mu) = \sum_{i}\ln(1 + e^{\beta\mu-\beta\epsilon_i})$$

$$= \sum_{i}\sum_{j}\frac{(-1)^{j-1}}{j}e^{j(\beta\mu-\beta\epsilon_i)} \quad (9.11)$$

Recall we defined $\tilde{z} = e^{\beta\mu}$ thus the coefficient of $e^{\beta\mu k}$ is $x_k = [(-1)^{k-1}/k]\sum_i e^{-k\beta\epsilon_i}$. Thus we can calculate the partition function

$$Q_A = \frac{1}{A}\sum_{k=1}^{A} k x_k Q_{A-k} \tag{9.12}$$

Here Q_0 is 1. When the expressions for x_k are used in the above equation, orbitals are given occupancies larger than 1 and then eliminated by subtraction. This can lead to severe round-off errors when applied to degenerate Fermi systems but will not affect the applications here.

As elementary examples: $x_1 = \sum_i e^{-\beta\epsilon_i}$ and $Q_1 = x_1$. Next $x_2 = -\frac{1}{2}\sum_i e^{-2\beta\epsilon_i}$ and

$$Q_2 = \frac{1}{2}[x_1 Q_1 + 2x_2 Q_0] = \frac{1}{2}\left[\sum_i e^{-\beta\epsilon_i}\sum_j e^{-\beta\epsilon_j} - e^{\sum_i e^{-2\beta\epsilon_i}}\right] \tag{9.13}$$

The second term in the equation above subtracts the double occupancies implied in the first term.

We will have to generalise to composites which will have two indices i and j, one giving the number of neutrons and the other giving the number of protons. Instead of x_k we will write $y_{0,1}^{[k]}$ meaning it is a composite with 0 neutron and 1 proton. The symbol k means it is obtained from the kth term in the expansion; $y_{0,1}^{[k]}$ will contribute to $x_{0,k}$.

If instead we had a boson, a deuteron for example, we would have

$$\ln[Z_{gr}(\beta, \mu_n, \mu_p)] = \sum_i -\ln(1 - e^{\beta\mu_n + \beta\mu_p}e^{-\beta\epsilon_i})$$

$$= \sum_i \sum_j \frac{1}{j}e^{j(\beta\mu_n + \beta\mu_p - \beta\epsilon_i)} \tag{9.14}$$

Thus in the case of deuterons (which would contribute to $x_{k,k}$) it is given by $(1/k)\sum_i e^{-k\beta\epsilon_i}$.

We can treat an assembly of protons, neutrons, deuterons, tritons, etc. The recursive relation if the dissociating system has N neutrons and Z protons is

$$Q_{N,Z} = \frac{1}{N} \sum_{i=1,N,j=0,Z} i x_{i,j} Q_{N-i,Z-j} \qquad (9.15)$$

So long as a partition function can be written as $Q_{N,Z} = \sum \prod \frac{x_{i,j}^{n_{i,j}}}{n_{i,j}!}$ with the condition that $\sum i n_{i,j} = N$ and $\sum j n_{i,j} = Z$ this will follow. In particular $x_{i,j}$ need not be the one-particle function of the composite with i neutrons and j protons.

In the formalism developed above, a given composite will contribute to many $x_{i,j}$. If the composite has i_1 neutrons and i_2 protons and its one-particle partition function at β is denoted by $\tilde{x}_{i_1,i_2}(\beta)$, then it contributes $y_{i_1,i_2}^{[1]} = \tilde{x}_{i_1,i_2}(\beta)$ to x_{i_1,i_2}, then $y_{i_1,i_2}^{[2]} = [\mp\frac{1}{2}]\tilde{x}_{i_1,i_2}(2\beta)$ to $x_{2\times i_1,2\times i_2}$ and in general, $y_{i_1,i_2}^{[k]} = [\frac{(\mp)^{(k-1)}}{k}]\tilde{x}_{i_1,i_2}(k\beta)$ to $x_{k\times i_1,k\times i_2}$.

One can evaluate the y factors by replacing sums over orbitals with integrations. For example, $y_{0,1}^{[k]} = \sum_i e^{-k\beta\epsilon_i}$ where the sum is replaced by $\int e^{-k\beta\epsilon} g(\epsilon) d\epsilon = 2(V/h^3)(2\pi m/k\beta)^{3/2}$. Here V is the available volume. We have included the proton spin degeneracy and m is the proton mass. For the deuteron $y_{1,1}^{[k]} = (1/k) \int e^{-k\beta\epsilon} g(\epsilon) d\epsilon$. This is $3 \times 2^{3/2}(V/h^3)(2\pi m/\beta)^{3/2}(e^{-k\beta E_b}/k^{5/2})$ where E_b is the binding energy of the deuteron. Finally we write down the average number of a composite with i_1 neutrons and i_2 protons:

$$\langle n_{i_1,i_2} \rangle = y_{i_1,i_2}^{[1]} \frac{Q_{N-i_1,Z-i_2}}{Q_{N,Z}} + 2y_{i_1,i_2}^{[2]} \frac{Q_{N-2i_1,Z-2i_2}}{Q_{N,Z}} + \cdots \qquad (9.16)$$

9.2 Results

We test the accuracy of the yields as calculated in the usual CTM (as practised in the previous chapters) by comparing with a calculation in which the complete theory of symmetrisation and antisymmetrisation is used. Subject only to the approximation that the summation over discrete states has been replaced by an integration over a density of states, the calculation is exact. The results are taken from [86], We take the dissociating system to have $Z = 25$ and $N = 25$. The lowest temperature considered is $3\,\text{MeV}$ (one might argue that at lower temperature a model of sequential decay is more appropriate). The highest temperature shown is 30 MeV. We take a freeze-out volume in which the composites can move freely, that is three times the volume of a normal nucleus with 50 nucleons. In addition to neutrons and protons we allow the possibility of composites. Excited states of composites were not allowed (they could have been included but the purpose of the exercise was to compare the two models, calculations without the inclusion of excited states were sufficient to reach conclusions). Spins and binding energies for deuteron, triton, ^{3}He and ^{4}He are taken from experiments. For higher mass composites the binding energy is taken from empirical mass formula. For fermions, spin 1/2 was assumed and for bosons spin 0 was assumed. For each Z we take $N = Z - 1, Z$ and $Z + 1$. We present in Table 9.1 average yields of protons, neutrons, deuterons, tritons, ^{3}He, ^{4}He and the sum of yields of all nuclei with charges greater than 12.

The approximation used in the main part of the text appears to be quite good.

Table 9.1: Comparison of calculations of average yields and E/A. "App" and "ext" represent the approximate and exact values. By exact we mean a calculation with proper symmetry. Sum over discrete orbitals in a box has been replaced by integration as is the usual practice. The temperature range of 3–6 MeV is of interest in many experiments. We also show results at 30 MeV.

Cal	p	n	d	t	^{3}He	^{4}He	$Z > 12$	T (MeV)	E/A (MeV)
app	0.307	0.032	0.050	0.007	0.054	0.679	0.945	3	−7.863
ext	0.306	0.031	0.051	0.007	0.053	0.696	0.945	3	−7.861
app	1.174	0.898	1.177	0.560	0.641	2.489	0.051	6	−4.117
ext	1.117	0.856	1.195	0.553	0.638	2.573	0.050	6	−4.135
app	4.127	3.955	4.812	2.099	2.052	1.985	0.000	12	4.401
ext	3.860	3.696	4.941	2.090	2.051	2.021	0.000	12	4.308
app	10.937	10.893	7.664	1.686	1.650	0.379	0.000	30	28.914
ext	10.512	10.468	7.885	1.732	1.696	0.395	0.000	30	28.844

Chapter 10

Towards Microscopic Models

10.1 Introduction

Although we have discussed a few models for heavy ion collisions, we have not calculated any nucleon–nucleon (nn) collisions. The assumption was many nn collisions take place and that leads to a thermal and chemical equilibrium. In that case details of collision do not matter; populations in different channels depend solely upon availability of phase space. But not all interesting details can be covered by the blanket of thermal and chemical equilibrium. For example in head on collision of two ions, nuclear material must get compressed: by how much? After compression the matter will expand out quickly. What is the time scale? We begin to look into such questions now [87].

10.2 Cascade model

In the simplest case, we can model a nucleus as a collection of N_A nucleons spread uniformly within a sphere of radius R. Upto R then the number of particles is proportional to $d(r^3)d(\cos\theta)d\phi$. We can assign the positions of N_A particles by going through the following loop N_A times:

$$r = R(x_1)^{1/3} \tag{10.1}$$

$$\cos(\theta) = 1 - 2x_2 \tag{10.2}$$

$$\phi = 2\pi x_3 \tag{10.3}$$

Here x_1, x_2, x_3 are random numbers. In Cartesian coordinates the particles are at $z = r \cos\theta$, $x = r \sin\theta \cos\phi$, $y = r \sin\theta \sin\phi$. The centre of the nucleus is at $\frac{1}{N_A} \sum_i \vec{r}_i$ which will be close to 0.

However, should one want one can also Monte Carlo positions of nucleons where the nucleus has a realistic diffuse surface. A model that has been used is a Myers distribution [88]:

$$\rho(r) = \rho_0 \left[1 - \left(1 + \frac{R}{a} \exp(-R/a) \frac{\sinh(r/a)}{r/a} \right) \right] \quad r < R \quad (10.4)$$

$$\rho(r) = \rho_0 \left[\frac{R}{a} \cosh(R/a) - \sinh(R/a) \right] \frac{e^{-r/a}}{r/a} r > R \qquad (10.5)$$

where $R = 1.18 A^{1/3}$ fm, $a = 1/\sqrt{2}$ fm.

This distribution is quite realistic and has the advantage that the equivalent sharp radius R is simply proportional to $A^{1/3}$ while the half density radius of a Fermi distribution does not have this simple proportionality. The other advantage is that no special normalisation is required. The total number is correct since

$$4\pi \int_0^\infty \rho(r) r^2 dr = \frac{4\pi}{3} \rho_0 R^3 \qquad (10.6)$$

Consider nucleus A start hitting nucleus B according to the simplest cascade model. The N_A nucleons are represented as N_A point particles according to our chosen prescription. Similarly N_B nucleons are represented as N_B point particles.

We may start by placing the centre of A at $X = b$, $Y = 0$, $Z = -D/2$ and the centre of B at $X = 0$, $Y = 0$ and $Z = +D/2$. D should be large enough that initially N_A nucleons are separated from N_B nucleons. We can start every nucleon in A with momentum $+p_z$ and every nucleon in B with momentum $-p_z$. Nucleons in A will invade the space of nucleons in B and nn collisions will start. We are describing a collision with impact parameter b.

The main task of a cascade program is to decide where and when particles collide. Divide collision time into small time intervals δt and examine all pairs of particles in each time interval to check whether they scatter. There are two conditions that need to be fulfilled.

The two nucleons must pass the point of closest approach in the time interval and the distance of closest approach must be less than $b_{max} = \sqrt{\sigma^t_{nn}(\sqrt{s})/\pi}$. Here $\sigma^t_{nn}(\sqrt{s})$ is the total cross-section for the nn CM energy $\sqrt{s}$. The time interval should be small enough that the probability of the same nucleon being scattered more than once in one time interval is very small. Usually δt is chosen between 0.2 and 0.5 fm/c.

Provided a collision occurs, the two particles can scatter elastically or inelastically. If the beam energy is 200 MeV/n or less, the inelastic channel can be suppressed and non-relativistic kinematics can be used with considerable simplification. In Appendix C we deal with the general case. The only inelastic channel is the pion channel. This is included as the Δ state of the nucleon. Here we will consider the following channels:

$$n + n \rightarrow n + n \text{ (a)}$$
$$n + n \rightarrow n + \Delta \text{ (b)}$$
$$n + \Delta \rightarrow n + n \text{ (c)}$$
$$n + \Delta \rightarrow n + \Delta \text{ (d)}$$
$$\Delta + \Delta \rightarrow \Delta + \Delta \text{ (e)} \tag{10.7}$$

Details of the cross-section parametrisation are given in Appendix C. The cross-sections of processes (a) and (b) are taken from experiments. Cross-section for channel (c) is obtained by detailed balance. Cross-sections for (d) and (e) are taken to be the same as (a). If two particles collide and if both elastic and inelastic channels are open a Monte carlo decision is made for the channel. Another Monte Carlo decision is taken to fix the angle of scattering. For elastic scattering differential scattering cross-section can be taken from experiments. Isotropic scattering is assumed in inelastic channel.

For each impact parameter and beam energy, the cascade model has to be run many times. Because particle positions and collisions are obtained by Monte Carlo sampling each run will produce different final results. We can regard each run as an event. The number of Δ's at the end will give the number of pions. It is of advantage to run all the events simultaneously. By averaging one can see the increase of density and the average number of collisions per unit time and inclusive cross-sections.

Appendix B is a guide for implementation of nn collisions at lower energies where Δ production is insignificant.

Chapter 11

Cascade Plus Mean Field Model

11.1 Introduction

Many very important features are missing in the cascade model
described in Chapter 10. For an event we considered N_A nucleons as
N_A point particles at different locations within a sphere of radius R.
The nucleons were set in motion in the z-direction by ascribing to
each nucleon a momentum p_z. For a stationary nucleus, p_z would be
zero and the nucleons would have no movements. This model has no
Fermi motion and no equation of state. One can ascribe to a nucleon
not only a position $\vec{r}$ but also, in a similar fashion, a momentum $\vec{p}$
within a Fermi sphere p_F but the nucleons will leak out of the sphere
of radius R very quickly unless there is a mean field which keeps the
nucleons together.

11.2 Cascade plus mean field

A transport model which incorporates both hard collisions and mean
field goes under several names: the Boltzmann–Uehling–Uhlenbeck
equation (BUU) [89,90], the Boltzmann–Nordheim equation [91] and
the Landau–Vlasov equation [92]. Nucleon positions imply a certain
density. We postulate that the potential a nucleon feels is a function
of the density. This is called Skyrme parametrisation. A potential
that has been widely used and explains many nuclear data very

successfully has the form

$$U(\rho) = A\left\{\frac{\rho}{\rho_0}\right\} + B\left\{\frac{\rho}{\rho_0}\right\}^{\sigma} \tag{11.1}$$

Here ρ is local density, $\rho_0 \approx 0.16\,\text{fm}^{-3}$ is normal nuclear density, A is attractive and B is repulsive. As we will see the choice of the values of A and B determines the equation of state of nuclear matter. How do we extend cascade model to define a density and also include collisions?

Consider the case of a nucleus with N_A nucleons hitting another nucleus with N_B nucleons at impact parameter b and a given beam energy. We already said that it is advantageous to have $\tilde{N}$ runs simultaneously. In the cascade model the different runs do not communicate with one another. Thus nucleus 1 hits nucleus $1'$, nucleus 2 hits nucleus $2', \ldots$, nucleus $\tilde{N}$ hits nucleus $\tilde{N}'$. Now we introduce communication between runs. What we were labelling as nucleons are now labelled as test particles. Thus we have $(N_A + N_B)\tilde{N}$ test particles. The density $\rho(\vec{r})$ is given by $\rho(\vec{r}) = N'/(\delta r)^3 \tilde{N}$ where N' is the number of test particles in the small volume $(\delta r)^3$ around the point $\vec{r}$ and $\tilde{N}$ already defined is the number of test particles per nucleon. As far as collisions go it is customary to segregate different runs but this is merely a computational trick. The test particles should hit each other with a cross-section $\sigma_{nn}/\tilde{N}$. By segregating runs we are able to use σ_{nn} and significantly reduce computation. As time progresses, test particles occasionally collide and in between collisions move clasically according to Hamilton's equations

$$\frac{d}{dt}\vec{p_i} = -\nabla_r U(\rho(\vec{r_i})) \tag{11.2}$$

$$\frac{d}{dt}\vec{r_i} = \vec{v_i} \tag{11.3}$$

We will return to collisions soon but first let us consider one nucleus N_A by itself. In the last chapter we considered a sphere of radius R and assuming uniform density assigned N_A positions for the nucleons inside the sphere using Monte Carlo sampling. The nucleons had positions but no momenta. Now we will assign momentum to each nucleon as well. In the ground state, nucleons will fill upto Fermi momentum p_F. We have $N_A = \frac{4}{h^3} \int \int d^3r d^3p$ where 4 takes account of spin–isospin degeneracy. This gives $\rho = \frac{16\pi}{3h^3} p_F^3$. From N_A and R, ρ is known which determines p_F. Just as for position, momentum can be picked by Monte Carlo sampling from a sphere of radius p_F for each nucleon. Thus now a nucleon has both a position and a momentum.

Let us get back to the description of 1 hitting 1′, 2 hitting 2′,..., $\tilde{N}$ hitting $\tilde{N}′$. That has not changed. The only difference is that in 1 and 1′ (and 2 and 2′ etc.), the nucleons were fixed in positions except for CM motions. Now the nucleons have also Fermi motion. As before nucleons become test particles, the number of test particles is $\tilde{N}$ per nucleon. Let us get back to two nuclei hitting each other. When two test particles collide they change from $(\vec{r_1}, \vec{p_1}), (\vec{r_2}, \vec{p_2})$ to $(\vec{r_1}, \vec{p_1'}), (\vec{r_2}, \vec{p_2'})$ (see Appendix B). If the phase space around $(\vec{r_1}, \vec{p_1'})$ and $(\vec{r_2}, \vec{p_2'})$ are essentially empty, the scattering will be allowed. If on the other hand they are nearly full, the scattering will be inhibited. This is called Pauli blocking. The implementation of this has varied [93]. Here we give details of one. To test the environment around $(\vec{r_1}, \vec{p_1'})$ we draw a sphere of radius R_P centering $\vec{r_1}$ and a sphere of radius P_P centering $\vec{p_1'}$ such that n test particles in this phase space volume implies complete filling. A reasonable value for n is 8. The number n cannot be taken to be very small because statistical fluctuations, inherent in Monte Carlo sampling, will be a source of error. It cannot be chosen too large because we need to sample the phase space filling close to $(\vec{r_1}, \vec{p_1'})$. Specifying n still does not fix R_P and P_P individually. A judicious choice is $R_P/P_P = R/p_F$ where R is the radius and p_F is the Fermi momentum of the static nucleus. Define $f_1' = n_1/(n-1)$ where n_1 is the number of test particles not including the particle at $(\vec{r_1}, \vec{p_1'})$. Similarly $f_2' = n_2/(n-1)$. The probability of scattering is taken to be $(1 - f_1')(1 - f_2')$ and this is Monte-Carlo'ed in the usual way.

What we have just described is a method first used in [90] to solve the BUU equation. This equation is

$$\frac{\partial f_1}{\partial t} + \vec{v} \cdot \vec{\nabla}_r f_1 - \vec{\nabla}_r U \cdot \vec{\nabla}_p f_1$$

$$= -\int \frac{d^3 p_2 d^3 p_1' d^3 p_2'}{(2\pi)^9} \sigma \vec{v}_{12}$$

$$\times \left\{ f_1 f_2 (1 - f_1')(1 - f_2') - f_1' f_2' (1 - f_1)(1 - f_2) \right\}$$

$$\times (2\pi)^3 \delta^3 (\vec{p} + \vec{p}_2 - \vec{p'}_1 - \vec{p'}_2) \tag{11.4}$$

The right hand side is the collision integral including Pauli blocking. We have included collisions using the cascade model and Pauli suppression has been included. The left hand side arises from mean-field theory. When the right hand side is set to zero, we have the Vlasov equation:

$$\frac{\partial f}{\partial t} + \frac{\vec{p}}{m} \cdot \vec{\nabla}_r f - \vec{\nabla}_r U \cdot \vec{\nabla}_p f = 0 \tag{11.5}$$

Bertsch has shown that the Vlasov equation can be obtained from the time-dependent Hartree–Fock theory after a few approximations and identifying $f(\vec{r}, \vec{p})$ as the Wigner transform for phase-space density [94]. For further details see [95]. A short introduction to Wigner transform is given in Appendix D. In our case we are replacing the continuous distribution $f(\vec{r}, \vec{p})$ by a large number of test particles. Thus we use

$$f(\vec{r}, \vec{p}) \approx \frac{1}{\tilde{N}} \sum_i \delta(\vec{r} - \vec{r_i}) \delta(\vec{p} - \vec{p_i}) \tag{11.6}$$

where $\vec{r_i}$ and $\vec{p_i}$ are coordinates of individual test particles. We choose those test particles i which are close to $\vec{r}$ and $\vec{p}$. We now get

$$\vec{\nabla}_r f = \frac{1}{\tilde{N}} \sum_i \delta'(\vec{r} - \vec{r_i}) \delta(\vec{p} - \vec{p_i}) \tag{11.7}$$

$$\vec{\nabla}_p f = \frac{1}{\tilde{N}} \sum_i \delta(\vec{r} - \vec{r_i})\delta'(\vec{p} - \vec{p_i}) \tag{11.8}$$

$$\frac{\partial f}{\partial t} = \sum_i \left\{ \delta'(\vec{r} - \vec{r_i})(\delta(\vec{p} - \vec{p_i})\frac{\partial \vec{r_i}}{\partial t} + \delta(\vec{r} - \vec{r_i})\delta'(\vec{p} - \vec{p_i})\frac{\partial \vec{p_i}}{\partial t} \right\} \tag{11.9}$$

It is now easy to see that Eq. (11.5) is satisfied if $\vec{r_i}, \vec{p_i}$ obey Hamilton's equations.

In the numerical implementation of the test particle method, the propagation of test particles is represented by

$$\vec{p}(t + \delta t) = \vec{p}(t) - \delta t \vec{\nabla}_r U \left(\vec{r}, t + \frac{1}{2}\delta t \right) \tag{11.10}$$

$$\vec{r}\left(t + \frac{1}{2}\delta t\right) = \vec{r}\left(t - \frac{1}{2}\delta t\right) + \delta t \frac{\vec{p}(t)}{m} \tag{11.11}$$

In usual prescription, BUU simulations are carried out in rectangular coordinate. Depending upon individual case we may work in a box which spans x from 0 to 80 fm, y from 0 to 80 fm and z from 0 to 100 fm. The volume of the box is divided into small cubes (say in d fm^3 cubes: often d is chosen to be 1). We need to label these cubes by three integers i, j, k. For example a cube whose x runs from 3 to 4 fm, y from 8 to 9 fm and z from 30 to 31 fm can be labelled $(4, 9, 31)$. Assuming calculations are being carried out in CM frame of the two nuclei one could start with the two nuclei almost touching each other near the centre of the box. We have already indicated how we could initialise 1 and 1', 2 and 2',..., $\tilde{N}$ and $\tilde{N}'$ with positions and momenta of $(N_A + N_B)\tilde{N}$ test particles. Assume the positions generated give positions at $t = 0$ and the momenta are those at $t = \delta t/2$. Then we can arrive at $\vec{r}(\delta t) = \vec{r}(0) + \delta t \vec{p}(\frac{\delta t}{2})/m$ and at $\vec{p}(\frac{3}{2}\delta t) = \vec{p}(\frac{1}{2}\delta t) - \delta t \vec{\nabla}_r U(\vec{r}, t)$ and thus we progress in time. The force field $\vec{\nabla}_r U$ is calculated as follows. When trying to determine the x-component of the force acting on a test particle which is in a cube labelled by i, j, k, we often smoothen the density by averaging over neighbouring cubes with some weighting factor. The x-component of the force on a particle in a cube is then taken to be $\frac{1}{2d}(U(\rho(i-1,j,k)) - U(\rho(i+1,j,k)))$ and similarly for other components. An improved way of calculating U and its derivative will be presented in Chapter 12.

With the introduction of the potential $U(\rho)$ we can begin to look at equation of state (EOS) of nuclear matter. Presence of $U(\rho)$ implies a potential energy density. This is given by $V(\rho) = \int_0^\rho U(\rho')d\rho'$. For $U(\rho)$ postulated in the beginning, $U(\rho) = A\left(\frac{\rho}{\rho_0}\right) + B\left(\frac{\rho}{\rho_0}\right)^\sigma$, the potential energy density $V(\rho)$ is $V(\rho) = \frac{A}{2}\frac{\rho^2}{\rho_0} + \frac{B}{\sigma+1}\frac{\rho^{\sigma+1}}{\rho_0^\sigma}$. One could take $V(\rho)$ as the more fundamental quantity and obtain an expression for $U(\rho)$ from $\frac{\partial V}{\partial \rho}$. One obtains

$$\frac{E}{N} = \frac{3}{10m}\left\{\frac{3h^3}{16\pi}\right\}^{2/3}\left\{\frac{\rho}{\rho_0}\right\}^{2/3} + \frac{A}{2}\left\{\frac{\rho}{\rho_0}\right\} + \frac{B}{\sigma+1}\left\{\frac{\rho}{\rho_0}\right\}^\sigma \quad (11.12)$$

The first term arises from kinetic energy, the next two from potential energy. Figure 11.1 shows the plot of E/N for two sets of parameters: (1) $A = -356\,\text{MeV}$, $B = 303\,\text{MeV}$, $\sigma = 7/6$ (this gives a soft EOS) and (2) $A = -124\,\text{MeV}$, $B = 70.5\,\text{MeV}$, $\sigma = 2$ (this gives a hard EOS). It is standard to quote a compressibility coefficient K defined by $9(\rho_0)^2\left\{\frac{d^2(E/N)}{d^2\rho}\right\}_{\rho_0}$. Here ρ_0 is the density where E/N

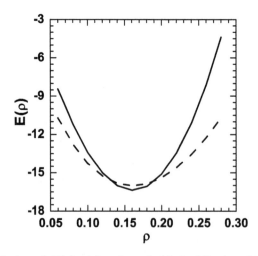

Fig. 11.1:   Variation of $E(\rho)$ with ρ for soft (dashed line) and hard (solid line) EOS.

minimises. The higher the value of K the harder it is to compress nuclear matter. Nuclear matter is a hypothetical very large nucleus where protons do not carry any charge. As is well known, proton charges limit the size of nuclei. Very dependable modelling postulates that nuclear matter has equilibrium density around $0.16\,\mathrm{fm}^{-3}$ and E/N about $-16\,\mathrm{MeV}$. Both sets of parameters we have used give these values (Fig. 11.1). The distinguishing feature is K which is $200\,\mathrm{MeV}$ for the soft EOS and $380\,\mathrm{MeV}$ for the hard EOS. Study of giant monopole vibrations in nuclei suggested a soft EOS and since in heavy ion collisions nuclei do get compressed, there was anticipation that heavy ion reactions would help in the extraction of the stiffness to compression. We will get to that later.

Figure 11.2 shows the variation of central density and number of collisions as a function of time obtained from BUU model calculation for ^{20}Ne on ^{20}Ne reaction at $400\,\mathrm{MeV/nucleon}$. Comparison of results from different BUU and QMD (described in Chapter 17) models can be found in [96].

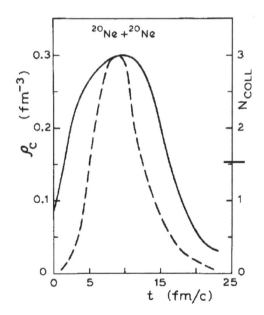

Fig. 11.2: Variation of central density ρ_c and number of collisions N_{coll} per $0.5\,\mathrm{fm/c}$ as a function of time for ^{20}Ne on ^{20}Ne reaction at $400\,\mathrm{MeV/nucleon}$.

Chapter 12

Lattice Hamiltonian Vlasov Method

Chapter 11 set up the basic structure of BUU. Depending upon the specific problem one wants to investigate, it can be added to or be modified. We will introduce several such modifications. In the first of such modifications, we will describe lattice Hamiltonian Vlasov (LHV) method whose objective is to improve the accuracy of the numerical procedure of Chapter 11. It can, in addition, use ions with proper diffuse surfaces for collisions. The ions introduced in Chapter 11 had sharp surfaces in the ground state (Appendix D). LHV was introduced by R. Lenk and V. Pandharipande [98].

In the transport model the test particles collide from time to time and in between propagate in a self-generated mean field (Vlasov propagation). The two-body collisions conserve energy and momentum exactly so any error related to energy and momentum non-conservation must come from the treatment of the mean field. In simulations of events with A nucleons using $A\tilde{N}$ test particles the test particle density

$$\rho_T(\vec{r}) = \frac{1}{\tilde{N}} \sum_{i=1}^{A\tilde{N}} \delta^3(\vec{r} - \vec{r_i}) \tag{12.1}$$

is not smooth. Most problems in Vlasov theory of nuclear physics arise from the need to smoothen $\rho_T(\vec{r})$.

In LHV one calculates the average density ρ_L at the sites of a three-dimensional cubic lattice:

$$\rho_L(\vec{r_\alpha}) = \sum_{i=1}^{A\tilde{N}} S(\vec{r_\alpha} - \vec{r_i}) \tag{12.2}$$

where α is the lattice index and $\vec{r_\alpha}$ is the position of the site α. The form factor used is

$$S(\vec{r}) = \frac{1}{\tilde{N}(nl)^6} g(x)g(y)g(z) \tag{12.3}$$

$$g(q) = (nl - |q|)\theta(nl - |q|) \tag{12.4}$$

Here l is the lattice spacing, θ is the Heaviside function and n is an integer which determines the range of S. Lenk and Pandharipande recommend $n = 2$ and $l = 1$ fm. But experience has shown that $n = 1$ and $l = 1$ fm give good enough energy and momentum conservation for applications that have been tried. It is seen that a test particle contributes to ρ_L at exactly $(2n)^3$ lattice sites and the movement of a test particle results in a continuous change at nearby lattice sites. This is absent in the easier method of Chapter 11 which makes accurate energy conservation difficult.

The form of $S(\vec{r})$ has the property

$$l^3 \sum_\alpha S(\vec{r_\alpha} - \vec{r}) = \frac{1}{\tilde{N}} \tag{12.5}$$

independent of $\vec{r}$ so that

$$l^3 \sum_\alpha \rho_L(\vec{r_\alpha}) = A \tag{12.6}$$

It is to be noted that the smoothened ρ_L and not ρ_T is the fundamental density in the Vlasov calculations since it determines $U(\vec{r})$ and thus the system dynamics.

In Chapter 11 we introduced the simple mean-field potential given in Eq. (11.1) and its potential energy density

$$v(\vec{r}) = \frac{A\rho_0}{2}\left\{\frac{\rho(\vec{r})}{\rho_0}\right\}^2 + \frac{B\rho_0}{\sigma + 1}\left\{\frac{\rho(\vec{r})}{\rho_0}\right\}^{\sigma+1} \tag{12.7}$$

In Chapter 11 we showed an approximate method of solving the transport equation. We can do an improved calculation for this mean-field potential and energy density with the LHV method. Lenk and Pandharipande also added to this mean-field potential another term $\frac{C}{\rho_0^{2/3}}\nabla_r^2\{\frac{\rho(\vec{r})}{\rho_0}\}$ and a corresponding potential energy density: $\frac{C\rho_0^{1/3}}{2}\{\frac{\rho(\vec{r})}{\rho_0}\}\nabla_r^2\{\frac{\rho(\vec{r})}{\rho_0}\}$. These added terms do not contribute to matter of uniform density but give a better description at the surface where the density drops from a constant to decreasing values.

So the expression of mean-field potential and potential energy density for finite nuclei can be written as

$$U(\rho) = A\left\{\frac{\rho}{\rho_0}\right\} + B\left\{\frac{\rho}{\rho_0}\right\}^\sigma + \frac{C}{\rho_0^{2/3}}\nabla_r^2\left\{\frac{\rho(\vec{r})}{\rho_0}\right\} \qquad (12.8)$$

$$v(\vec{r}) = \frac{A\rho_0}{2}\left\{\frac{\rho(\vec{r})}{\rho_0}\right\}^2 + \frac{B\rho_0}{\sigma+1}\left\{\frac{\rho(\vec{r})}{\rho_0}\right\}^{\sigma+1} + \frac{C\rho_0^{1/3}}{2}\left\{\frac{\rho(\vec{r})}{\rho_0}\right\}\nabla_r^2\left\{\frac{\rho(\vec{r})}{\rho_0}\right\}$$
$$(12.9)$$

The following values of force parameters give reasonable numbers for ground state properties of finite nuclei in Thomas–Fermi theory (details are given in Appendix E). The values of the constants are $A = -356.8\,\text{MeV}$, $B = 303.9\,\text{MeV}$, $\sigma = 7/6$, $\rho_0 = 0.16\,\text{fm}^{-3}$ and $C = -6.5\,\text{MeV}$. In LHV the velocities of test particles evolve as before: $\frac{d\vec{r}_i}{dt} = \frac{\vec{p}_i}{m}$. Next we need a formula to compute how the momentum of a test particle changes. For this let us find out how the total energy changes as a test particle moves. We have discretised the configuration into boxes of size l^3 and the total interaction energy is $V = l^3\sum_\alpha v(\vec{r}_\alpha)$. As the test particle i makes an infinitesimal move the interaction energies in only some of the cubes change, not in others. We need to sum over only those α's. From one of this, the contribution to force is $-U(\rho_\alpha)\vec{\nabla}_i\tilde{S}(r_\alpha - \vec{r}_i)$. Here $\tilde{S}(\vec{r})$ is $\frac{1}{n^6 l^3}g(x)g(y)g(z)$ where $g(x), g(y), g(z)$ are the same as written before. Adding up contributions from all the contributing lattices one gets the net force and hence the change in momentum of the test particle i.

In Chapters 13, 14 and 15 we will show examples where LHV is used.

Chapter 13

Hybrid Model for Central Collision around Fermi Energy Domain

Central collision fragmentation reactions around Fermi energy domain are extensively used for producing neutron rich isotopes and for studying nuclear liquid–gas phase transition. We have developed a hybrid model for explaining multifragmentation reaction around the Fermi energy domain in order to treat central collision [99]. Initially the excitation of the colliding system is calculated by using dynamical Boltzmann–Uehling–Uhlenbeck (BUU) approach with proper consideration of pre-equilibrium emission. Then the fragmentation of this excited system is calculated by canonical thermodynamical model (CTM). The decay of excited fragments produced in multifragmentation stage is calculated by an evaporation model based on the Weisskopf theory. Different observables like charge distribution, largest cluster distribution etc. are calculated using this hybrid model for $^{129}Xe + ^{119}Sn$ reaction at different projectile energies and compared with experimental data [100]. The idea of setting the initial conditions for a statistical model from a dynamical model is of course not new; see for example Barz *et al.* [101]. In many statistical models of multifragmentation (SMM), the initial conditions are fixed by some measured data. In our hybrid model the initial conditions for the thermodynamical model are set up almost entirely by the transport model calculation.

   The concept of temperature is quite familiar in heavy ion collision and it is a better observable (compared to energy) for studying liquid–gas phase transition. One standard way of extracting temperatures is the Albergo formula [44], where temperature is calculated from the measured isotopic yields (i.e. cold fragments). Another common technique for obtaining temperature is to measure the kinetic energy spectra of emitted particles. But in this method, the effect of sequential decay from higher energy states, Fermi motion, pre-equilibrium emission etc. complicates the scenario of temperature measurement. Our hybrid model calculation is an alternative method for deducing the freeze-out temperature and it bypasses all such problems.

13.1 Basics of the dynamical model calculation for initial condition

We start our calculation when two nuclei in their respective ground states approach each other with specified velocities. The mean-field potential energy density is taken from Lenk–Pandharipande [98] and is given in details in Chapter 12 (Eq. (12.9)). We first construct the Thomas–Fermi solutions for ground states [102]. The Thomas–Fermi method is described in Appendix E. The Thomas–Fermi phase space distribution will then be modeled by choosing test particles with appropriate positions and momenta using Monte Carlo. Each nucleon is represented by 100 test particles ($\tilde{N} = 100$). Then we begin BUU model calculation to get the excitation of the fragmenting system. In the CM frame the test particles of the projectile and the target nuclei (in their Thomas–Fermi ground state) are boosted towards each other. The test particles move in a mean field $U(\rho(\vec{r}))$ (generated by the potential energy density) and will occasionally suffer two-body collisions when two of them pass close to each other and the collision is not blocked by Pauli principle. The mean field propagation is done using the lattice Hamiltonian Vlasov method described in Chapter 12 which conserves energy and momentum very accurately [98]. Two-body collisions are calculated as in Appendix B. Positions and momenta of the test particles are updated after each time steps (Δt) by the equations as in Eqs. (11.10) and (11.11).

13.2   Excitation energy determination

We can calculate the excitation energy (E^*) from projectile beam energy (E_{beam}) by direct kinematics by assuming that the projectile and the target fuse together. In that case the excitation energy is $E^* = A_p E_{beam}/(A_p + A_t)$ where A_p and A_t are projectile and target masses respectively. This value is too high as a measure of the excitation energy of the system which multifragments. Pre-equilibrium particles which are not part of the multifragmenting system carry off a significant part of the energy.

To get a better measure of excitation of the fragmenting system we need to do a BUU calculation where the pre-equilibrium particles can be identified and can be taken out to calculate excitation energy per nucleon. We exemplify our method with central collision reactions ^{129}Xe $+ {}^{119}$Sn at projectile beam energy 45 MeV/nucleon. Initially the centres of ^{129}Xe and ^{119}Sn are kept at (100 fm, 100 fm, 90 fm) and (100 fm, 100 fm, 110 fm) respectively and they are boosted towards each other along z-direction. Figure 13.1 shows the test particles at $t = 0$ fm/c (when the nuclei are separate), 75 fm/c (the time when violent collisions occur) and 200 fm/c (almost all collisions are completed). From the figure it is clear that for $t = 200$ fm/c some test particles are far from the central dense region. These fit the category of pre-equilibrium emission. In different multifragmentation

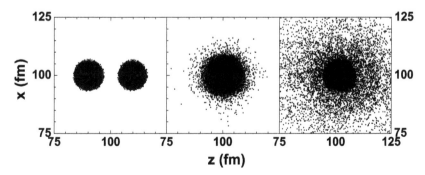

Fig. 13.1:   Evolution of test particles in CM frame for 45 MeV/nucleon ^{129}Xe on ^{119}Sn reaction at 0 fm/c (left panel), 75 fm/c (middle panel) and 200 fm/c (right panel).

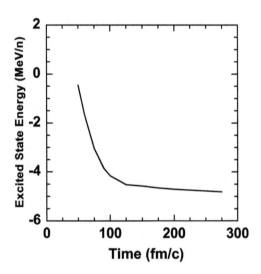

Fig. 13.2: Variation of energy of the central dense region (containing 80% of total test particles) with time obtained from dynamical BUU calculation for ^{129}Xe on ^{119}Sn reaction at 45 MeV/nucleon.

experiments, it is observed that after pre-equilibrium emission, 75% to 80% of the total mass creates the fragmenting system [103, 104]. Hence we choose the test particles which create 80% of the total mass (i.e. $A_0 = 198$) from the most central dense region. Knowing the momenta of selected test particles the kinetic energy is calculated and from the positions of these selected test particles the potential energy is calculated using Eq. (12.9). By adding kinetic and potential energies the energy of the fragmenting system is obtained. Figure 13.2 shows the variation of excited state energy of the central dense region (i.e. 80% of the total test particles) with time. Here total energy is always constant but as time progresses, pre-equilibrium particles, having high kinetic energy, escape from the central dense region, therefore the energy of the central dense region decreases. It is clear that after $t = 100$ fm/c, the energy becomes independent of time. Hence, we can stop BUU calculation at any time after $t = 100$ fm/c and consider the corresponding energy as excited state energy. To get the excitation we need to know the ground state of the fragmenting system. For this we use the Thomas–Fermi method for a spherical nucleus of mass $A = 198$ (80% of ^{129}Xe $+ {}^{119}$Sn mass). Subtracting ground state energy from the calculated energy above, the excitation energy is obtained.

13.3   Computations with the statistical model: Extraction of temperature

We have described above how from BUU we extract the mass, charge and the excitation energy of the fragmenting system. Our next task is to obtain the freeze-out temperature. The canonical thermodynamic model (CTM) can be used to calculate average excitation per nucleon for a given temperature, charge and mass. Getting an excitation energy for a given temperature, mass and charge is described in detail in Chapter 3. We do the exploration for each beam energy. We will not repeat the formulae of CTM here but just mention that apart from neutrons and protons the following composites are included in CTM breakup. We include deuteron, triton, ^{3}He, ^{4}He and for heavier nuclei we include a ridge along the line of stability. The composites that follow from CTM will further decay by evaporation described in Chapter 3.

13.4   Results

We have done calculations for the same ^{129}Xe $+ {}^{119}$Sn pair for projectile beam energies 32, 39, 45 and 50 MeV/nucleon. In each case, we have stopped the time evolution at $t = 200$ fm/c, and selected 80% of the total mass from central dense region for calculating the excited state energy. Then subtracting the ground state energy the excitation is obtained. The variation of calculated excitation energy with projectile beam energy is shown in the left panel of Fig. 13.3. From this excitation energy we find out the corresponding freeze-out temperature (shown in right panel of Fig. 13.3). Thus the freeze-out temperature for a given beam energy is obtained.

To check the accuracy of our model, we have compared the theoretical results with experimental data. Figure 13.4 shows the comparison of charge distribution at projectile beam energies 32, 39,

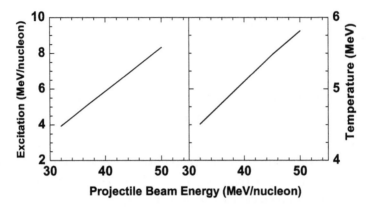

Fig. 13.3: Variation of excitation energy per nucleon (left panel) and temperature (right panel) with projectile beam energy per nucleon.

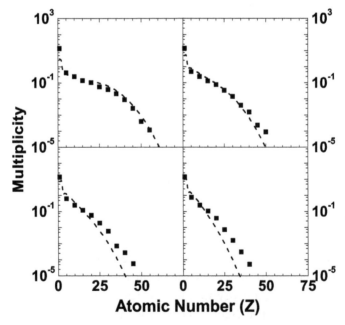

Fig. 13.4: Theoretical charge distribution (dotted lines) for ^{129}Xe on ^{119}Sn reaction at 32 MeV/nucleon (top left panel), 39 MeV/nucleon (top right panel), 45 MeV/nucleon (bottom left panel) and 50 MeV/nucleon (bottom right panel). The experimental data are shown by black squares.

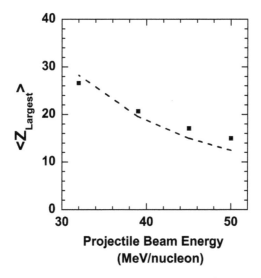

Fig. 13.5:   Variation of average size of largest cluster with projectile beam energy obtained from hybrid model calculation (dotted line) for ^{129}Xe on ^{119}Sn reaction. The experimental data are shown by squares.

45 and 50 MeV/nucleon. With the increase of energy (i.e. increase of temperature), there is more fragmentation, therefore multiplicities of higher fragments gradually decrease. The variation of average charge of largest cluster $\langle Z_{Largest} \rangle$ with projectile beam energy is shown in Fig. 13.5. In each case nice agreement between theoretical result and experimental data is obtained.

13.5   Discussions

This work complements the work we did where we fitted the data obtained from the decay of projectile-like fragments at energies in the limiting fragmentation region described in Chapter 5. There we fitted the data using CTM with an assumed temperature profile first and later [52] we showed that the temperature profile is obtainable from BUU calculations. Our present calculations are not prohibitively computer intensive. One virtue of these calculations is that used potential leads to reasonable values of binding energy of finite nuclei (even in Thomas–Fermi approximation) and realistic diffuse surface without having to supplement the zero range Skyrme interaction with a finite range interaction. Vlasov propagation for large nuclei when finite range interaction is present is very computer intensive. The other pleasing aspect is that the lattice Hamiltonian Vlasov method gives remarkable accuracy in total energy and total momentum conservation in these calculations.

For fragmenting system, we adopted the value of 80% of total mass from the experimental papers quoted in our paper. But our results (see Fig. 13.5) show that this was a reasonable choice. We show a plot of $\langle Z_{Largest} \rangle$ which agrees fairly well with data. Now $\langle Z_{Largest} \rangle$ depends upon the size of the fragmenting system as well as the temperature of the fragmenting system. The larger the fragmenting system, the larger is the $\langle Z_{Largest} \rangle$. The higher the temperature, the smaller is $\langle Z_{Largest} \rangle$. Now the temperature also depends upon what percentage of nucleons are left out as pre-equilibrium particles. The value 80% we choose gives a combination of temperature and fragmenting mass that seems to be just about right. One could do a detailed best "fit" but this was not attempted.

What we presented in this work did not involve any radial flow. One reason is that the collision energy being only about $50\,\mathrm{MeV/nucleon}$, the initial compression is small so any radial flow must also be small. The best signature for radial flow will be in the velocity distribution but we are only calculating multiplicity distribution. Neither CTM nor SMM can incorporate radial flow easily but in lattice gas model, where flow is easily incorporated, it was found that even for significant radial flow, multiplicity distributions are hardly affected [105].

Chapter 14

Projectile Fragmentation Combining BUU and CTM

14.1 Introduction

In Chapter 5 a model for projectile fragmentation has been proposed and different important observables have been studied from this model. The theoretical results obtained from this model have been compared with many experimental data with good success. The initial stage of the model is abrasion, where the PLF mass is determined from geometrical calculation. The PLF will have an excitation energy. It is conjectured that this will depend upon the relative size of the PLF with respect to the projectile, i.e., on (A_s/A_p) where A_s is the size of the PLF and A_p is the size of the whole projectile. Instead of excitation energy the concept of freeze-out temperature T is used and this temperature is not calculated, rather it is parameterised with the help of experimental data and used for further multifragmentation stage calculation (by using canonical thermodynamical model).

This chapter deals with the calculation of excitation and mass of the PLF directly from microscopic Boltzmann–Uehling–Uhlenbeck (BUU) transport model described in Chapter 11. Then by knowing PLF mass and excitation, the temperature is determined by using canonical thermodynamical model (CTM).

14.2 Searching initial conditions

The transport model calculation is started by choosing an impact parameter and boost the test particles of one nucleus with appropriate velocities in its Thomas–Fermi ground state (described in Appendix E) towards the test particles of the other nucleus, also in its ground state. First, it was chosen to study ^{58}Ni on ^{9}Be reaction with beam energy 140 MeV/nucleon which was experimentally investigated at Michigan State University (MSU) and also studied theoretically in Chapter 5. The transport calculations are done in a $25 \times 25 \times 31$ fm^3 box and for calculating the mean field in addition to zero-range Skyrme interaction, contribution from finite range Yukawa interaction was included, which includes the effect of diffuse nuclear surfaces. Therefore the form of the mean-field potential is given by

$$U(\vec{r}) = A'' \rho(\vec{r}) + B'' \rho^{\sigma}(\vec{r}) + \int u_y(\vec{r}, \vec{r'}) \rho(\vec{r'}) d^3 r' \qquad (14.1)$$

The constants are taken as, $A'' = -1563.6$ MeVfm^3, $B'' = 2805.3$ MeVfm$^{7/6}$, $\sigma = 7/6$, $V_0 = -668.65$ MeV and finite range parameter for Yukawa is 0.45979 fm. One can also replace Skyrme + Yukawa interaction potential by the potential given in Eq. (12.8). The configuration space is divided into 1 fm^3 boxes. It is useful to work in the projectile frame. Initially the projectile and target test particles are centered at (13 fm, 13 fm, 20 fm) and $(13 + b$fm, 13 fm, 27 fm) respectively and the target test particles are moving with the beam velocity in the negative z-direction. In this energy domain Vlasov propagation is treated non-relativistically.

We exemplify our method [52] with collision at impact parameter $b = 4$ fm. Figure 14.1 shows the test particles at $t = 0$ fm/c (when the nuclei are separate), $t = 10$ fm/c, $t = 25$ fm/c and $t = 50$ fm/c (Be has traversed the original Ni nucleus). The calculation was started with the centre of Ni at 20 fm; at the end a large blob remains centered at 20 fm. Clearly this is the PLF. However for a quantitative estimate of the mass of the PLF and its energy requires further analysis. This type of analysis was done for each pair of ions and at each impact

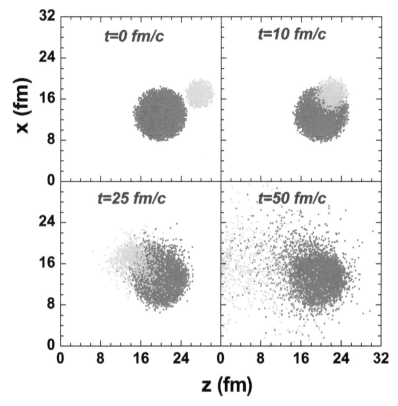

Fig. 14.1: Time evolution of ^{58}Ni (red) and ^{9}Be (green) test particles for 140 MeV/nucleon at an impact parameter $b = 4$ fm.

parameter and details vary from case to case. This is exemplified for $b = 4$ fm only.

For the analysis, it is convenient to introduce z-projection of density which is defined as

$$\rho_z(z) = \sum_{l,m} \rho_L(\vec{r}_\alpha) = \sum_{l,m} \sum_{i=1}^{AN_{test}} S(\vec{r}_\alpha - \vec{r}_i) \qquad (14.2)$$

where $\rho_L(\vec{r}_\alpha)$ is the density at a lattice point $\vec{r}_\alpha$ given in Eq. (12.2). The symbol α stands for values of the coordinates of the lattice point $\alpha = (x_l, y_m, z_n)$. We will often, for a fixed value of z_n, sum over

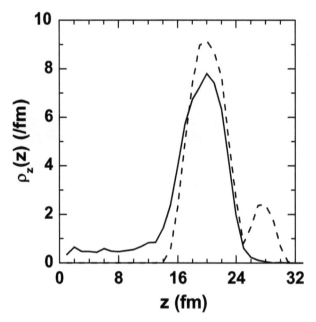

Fig. 14.2: Variation of $\rho_z(z)$ with z at $t = 0\,\text{fm/c}$ (dashed line) and $50\,\text{fm/c}$ (solid line) for $140\,\text{MeV/nucleon}$ ^{58}Ni on ^9Be reaction studied at an impact parameter $b = 4\,\text{fm}$.

x_l, y_m. For example $\sum_{l,m} \sum_{i=1}^{ANtest} S((x_l, y_m, z_n) - \vec{r}_i)$ will be denoted by $\rho_z(z_n)$. Similarly for kinetic energy or total energy density: $T(z_n)$ or $E_T(z_n)$. Similarly for $p_z c(z_n)$ etc.

In Fig. 14.2, $\rho_z(z)$ is plotted as a function of z at $t = 0$ (when the nuclei start to approach each other) and at $t = 50\,\text{fm/c}$ (when ^9Be has traversed ^{58}Ni).

The kinetic energy density is defined as

$$T_L(\vec{r}_\alpha) = \sum_{i=1}^{ANtest} T_i S(\vec{r}_\alpha - \vec{r}_i) \tag{14.3}$$

where T_i is the kinetic energy of the ith test particle. Therefore the kinetic energy per nucleon is:

$$\mu(z) = \frac{T_z(z)}{\rho_z(z)} = \frac{\sum_{l,m}(T_z)_L(\vec{r}_\alpha)}{\sum_{l,m} \rho_L(\vec{r}_\alpha)} \tag{14.4}$$

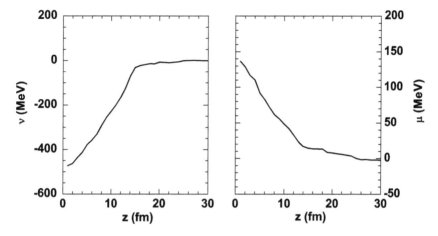

Fig. 14.3: Variation of momentum per nucleon $\nu(z)$ (left panel) and total energy per nucleon $\mu(z)$ (right panel) for $140\,\text{MeV/n}$ ^{58}Ni on ^{9}Be reaction at an impact parameter $b = 4\,\text{fm}$ studied at $t = 50\,\text{fm/c}$.

Similarly one can introduce a density for the zth component of momentum (actually $p_z c$ is used rather than p_z, where c is speed of light in vacuum)

$$(p_z c)_L(\vec{r}_\alpha) = \sum_{i=1}^{AN_{test}} (p_z c)_i S(\vec{r}_\alpha - \vec{r}_i) \qquad (14.5)$$

Therefore, $p_z(z)c$ per nucleon can be expressed as:

$$\nu(z) = \frac{p_z(z)c}{\rho_z(z)} = \frac{\sum_{l,m}(p_z c)_L(\vec{r}_\alpha)}{\sum_{l,m}\rho_L(\vec{r}_\alpha)} \qquad (14.6)$$

Figure 14.3 adds more details to the situation at $50\,\text{fm/c}$ where kinetic energy per nucleon (μ) and z-component of momentum per nucleon (ν) is plotted as a function of z. At far right, μ and ν are very small which indicates PLF regions (since the calculation is done in the projectile frame therefore the PLF have very low z-component of momentum and kinetic energy). Progressively towards left one has the participant zone characterised by a higher μ and lower value of ν. Closer to the left edge one has target spectators.

In order to specify the mass number and energy per nucleon of the PLF one needs to specify which test particles belong to the

PLF and which to the rest (participant and target spectators). The configuration box stretches from $z = 0$ to $z = 31$ fm. If all test particles in this range are included, one gets the full system with the total particle number 67 $(58 + 9)$ and the total energy of beam plus projectile in the projectile frame. Consider a wall at $z = 0$ fm and pulling the wall to the right. If the wall is pulled, the test particles positioned on the left of the wall are left out. With the test particles to the right of the wall one can compute the number of nucleons. The energy per nucleon right of the wall is labelled as E_{wr}. The number of particles goes down and initially the energy per nucleon E_{wr} will go down also as the target spectators and then the participants are being left out. At some point one enters the PLF and if pulls a bit further, part of the PLF is cut off and a non-optimum shape is formed. So the energy per nucleon E_{wr} will rise. The situation is shown in Fig. 14.4. The point which produces this minimum is a reference point. The test particles to the right are taken to belong to PLF; those to the left are taken to represent the participants and target spectators. Not surprisingly, this point is in the neighborhood where both μ and ν flatten out. The energy per nucleon E_{wr} at the reference point is the PLF excited state energy.

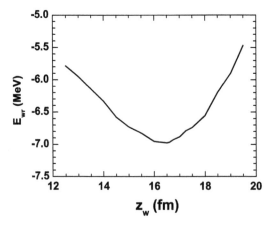

Fig. 14.4:   Energy per nucleon (E_{wr}) of the test particles remains right side of the separation (z) for 140 MeV/nucleon ^{58}Ni on ^{9}Be reaction at an impact parameter $b = 4$ fm studied at $t = 50$ fm/c.

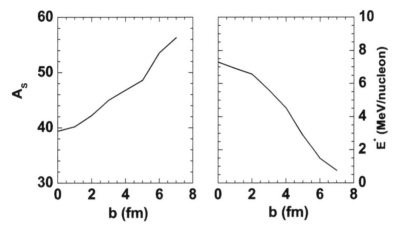

Fig. 14.5: Variation of PLF mass (A_s) (left panel) and its excitation per nucleon (E^*) (right panel) with impact parameter obtained from BUU calculation for ^{58}Ni $+ ^9$Be reaction at 140 MeV/nucleon.

Now, to calculate the excitation, ground state energy of the PLF is required. This is calculated from the Thomas–Fermi method described in Appendix E. Subtracting ground state energy from excited state energy excitation is obtained. Same procedure is followed for each impact parameter. For ^{58}Ni on ^{9}Be reaction at 140 MeV/nucleon, the variation of PLF mass (A_s) and excitation per nucleon (E^*) with impact parameter, obtained from this calculation, is shown in Fig. 14.5. As expected, with the increase of impact parameter, the total amount of mass which are driven out from the original projectile decreases, hence PLF mass increases. Also with the decrease of centrality, the deformation of the PLF decreases, therefore PLF excitation decreases.

The canonical thermodynamic model (CTM) can be used to calculate average excitation per nucleon for a given temperature, charge and mass. Getting an excitation energy for a given temperature, mass and charge is described in detail in Chapter 3. We do the exploration for each impact parameter. The variation of PLF temperature obtained from the above method and its comparison with universal temperature profile given in Eq. (5.3) is shown in Fig. 14.6.

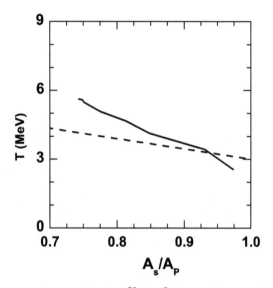

Fig. 14.6: Temperature profile for ^{58}Ni $+$ ^{9}Be reaction at $140\,$MeV/nucleon obtained from BUU model calculation (solid line) compared with that calculated from general formula given in Eq. (5.3) (dashed line).

14.3   Further results

Vlasov propagation with Skyrme plus Yukawa for large ion collisions is not practical. Given nuclear densities on lattice points, one is required to generate the potential which arises from the Yukawa interaction. Standard methods require iterative procedures involving matrices. In the case of Ni on Be, in the early times of the collision, the matrices are of the order of 1000 by 1000; as the system expands the matrices grow in size reaching about 7000 by 7000 at $t = 50\,$fm/c. If we want to do large systems (Sn on Sn for example) very large computing efforts are required.

To treat large but finite systems we use the mean-field Hamiltonian Lenk and Pandharipande devised for finite nuclei (described in Chapter 12). The mean field involves not only the local density but also the derivative of local density up to second order. The derivative terms do not affect nuclear matter properties but in a finite

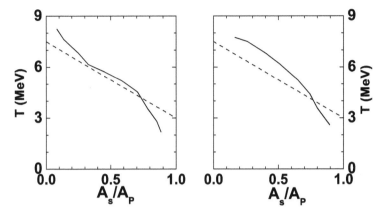

Fig. 14.7: Temperature profile for ^{58}Ni + ^{181}Ta reaction at 140 MeV/nucleon (left panel) and ^{124}Sn + ^{119}Sn reaction at 600 MeV/nucleon (right panel) obtained from BUU model calculation (solid lines) compared with that calculated from general formula given in Eq. (5.3) (dashed lines).

system it produces quite realistic diffuse surfaces and liquid-drop binding energies. By using the same procedure described above, the temperature profiles of ^{124}Sn + ^{119}Sn reaction at 600 MeV/n and ^{58}Ni + ^{181}Ta reaction at 140 MeV/n were determined and compared with universal temperature profile. This is shown in Fig. 14.7.

Chapter 15

A Model for Fluctuation in BUU

The BUU approach as advertised in Chapter 11 will not give clusters directly. It only describes average behaviour whereas clusters result from fluctuations. Clusters appear because in a given event a certain number of nucleons can appear in a small volume of the phase space. The average of all such fluctuations is still a smooth phase space density and this average phase space density is the final output of the BUU approach.

In intermediate energy heavy ion collision theory a model for fluctuations was first proposed by Bauer *et al.* [106]. We describe the model here. In the past event-by-event simulation in this model was limited to about mass number 30 on 30 [107]. The problem was a practical one. It required a large computing effort. It is now known that with a slight reformulation without changing any physics or numerical accuracy one can very significantly reduce the execution time and can handle much larger systems. Computation becomes as short as ordinary BUU calculations. It is instructive to handle large systems (finite number effects often hide important bulk effects) and more importantly fragmentation must be investigated over an energy range to unravel many interesting effects.

There are many microscopic models for multifragmentation. There are some which can be labelled as "quantum molecular dynamics" type [108,109]. These are different in spirit from the model described here. Closer in spirit yet quite distinct are some studies based on a Langevin model [110–114]. The literature in Langevin model is huge. We have mentioned only a few.

15.1 The prescription

Let us assume we are considering central collisions of two nearly equal ions A and B with particle numbers N_A and N_B. As mentioned in Chapter 11, it is convenient to run several events simultaneously. Let us denote the number of runs by $\tilde{N}$. In cascade different runs do not communicate with each other. Thus 1 hits $1'$, 2 hits $2', \dots, \tilde{N}$ hits $\tilde{N}'$. In BUU we introduced communication between runs. What we were calling nucleons we now call test particles (abbreviated from now on as tp). The density $\rho(\vec{r})$ is given by $n/(\delta r)^3 \tilde{N}$ where n is the number of tp's in a small volume $(\delta r)^3$. As far as collisions go, in usual applications of BUU one still segregates different runs but this is a computational shortcut. By segregating collisions one is able to use σ_{nn}, the nucleon–nucleon cross-section, and reduce computation. If one uses collisions between all tp's the collision cross-section would have to be reduced to $\sigma_{nn}/\tilde{N}$ and computation would be much longer.

To obtain multifragmentation with multiplicity n_a as a function of a where a is the mass number of the composite, Bauer *et al.* made the following prescription [106]. Now all tp's are allowed to collide with one another with a cross-section $\sigma_{nn}/\tilde{N}$. Collisions are further suppressed by a factor $\tilde{N}$ but when two tp's collide not only these two but also $2(\tilde{N}-1)$ tp's closest to these two in phase space change momenta. Physically this represents two actual particles colliding. When collisions cease, we have an event. A second event needs a new Monte Carlo of tp's and then the evolution in time.

We can use the same shortcut technique that is utilised in the usual BUU. There one uses for collisions 1 on $1'$, 2 on $2'$ etc. and σ_{nn}. To generate an event we do, for example, collisions in 1 on $1'$, but if two tp's successfully collide not only these two but $2(\tilde{N}-1)$ tp's closest to these two in phase space are modified. Let us label the two tp's that collided successfully as i and j. Label the tp's that will be modified along with i as i_s, s going from 1 to $\tilde{N}-1$.

Similarly the tp's that will be modified along with j are labeled j_s. The tp j_s's cannot include i or any i_s tp's. The phase space distance between tp i and tp i_s is taken to be $(\vec{r_i} - \vec{r_{is}})^2/R^2 + (\vec{p_i} - \vec{p_{is}})^2/p_F^2$ where R is the normal radius of A and p_F the Fermi momentum. Similarly for other phase space distances. All tp's labelled by i_s will be given the same momentum change $\vec{\Delta p}$ as suffered by tp i in the collision and all tp j_s get an additional momentum $-\vec{\Delta p}$ suffered by tp j.

This procedure will conserve momentum but not energy. With a little bit extra work both energy and momentum can be conserved. Define $\langle \vec{p_i} \rangle = \frac{\sum \vec{p_{is}}}{N}$; similarly $\langle \vec{p_j} \rangle$. One then considers a collision between $\langle \vec{p_i} \rangle$ and $\langle \vec{p_j} \rangle$ and obtain a $\vec{\Delta p}$ for $\langle \vec{p_i} \rangle$ and $-\vec{\Delta p}$ for $\langle \vec{p_j} \rangle$. This $\vec{\Delta p}$ is added to all $\vec{p_{is}}$ and $-\vec{\Delta p}$ to all $\vec{p_{js}}$. This conserves both energy and momentum as one progresses in time.

For the next event we can go back to the beginning and start the same procedure with 2 on 2'. Or we can start with new Monte Carlo sampling.

For mass 40 on 40 we compare this new shortcut results with the results obtained with the original method [106]. This is shown in Fig. 15.1. The agreement between the two calculations is remarkable.

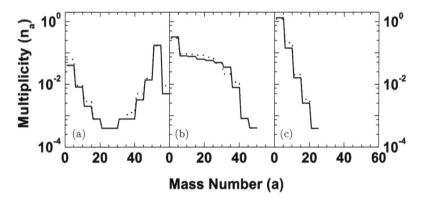

Fig. 15.1:   Comparison of mass distribution calculated according to the prescription of [106] (dotted lines) and the present work (solid lines). The average value of 5 mass units is shown. The cases are for central collision of mass 40 on mass 40 for different beam energies: (a) 25, (b) 50 and (c) 100 MeV/nucleon. 500 events were chosen at each energy.

The saving in computational time is so good that one can try bigger calculations and map an energy range to study how results change with beam energy.

15.2   Some details of the simulations

We provide some details of the calculation. We use the lattice Hamiltonian Vlasov method [98]. The potential energy density is given by Eq. (12.9). The last term in the right hand side of Eq. (12.9) gives rise to surface energy in finite nuclei. That favours the formation of larger composites, for example, the occurrence of a nucleus A over the formation of two nuclei of $A/2$ nucleons. Entropy favours break up.

The calculations were done in a $200 \times 200 \times 200\,\mathrm{fm}^3$ box. The configuration space was divided into $1\,\mathrm{fm}^3$ boxes. For result shown here the code was run from $t = 0\,\mathrm{fm/c}$ to $t = 200\,\mathrm{fm/c}$. Positions and momenta were updated every $0.3\,\mathrm{fm/c}$. For details of nucleon–nucleon collisions see Appendix B. The number of test particles $\tilde{N}$ per nucleon was set at 100. Once the two-body collisions become very rare, contiguous boxes with tp's that propagate together for a long time are considered to be part of the same cluster. The contiguous boxes have one common surface and the nuclear density exceeds a minimum value (d_{min}). Different d_{min} values were investigated (from 0.002 to $0.02\,\mathrm{fm}^{-3}$). Fragment multiplicity does not change very much with d_{min}. The value $0.01\,\mathrm{fm}^{-3}$ was used in what is shown in the results.

15.3   Results

In Fig. 15.2 we show plots of multiplicity against mass number a for 120 on 120. Four beam energies are shown. For each energy 1000 events were taken. We show results of averages of five consecutive mass numbers. The important features are: at low energy ($50\,\mathrm{MeV/nucleon}$) the multiplicity first falls with mass number a, reaches a minimum, then rises, reaches a maximum

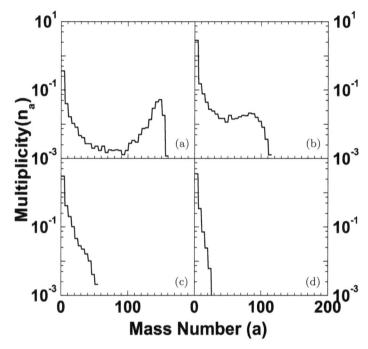

Fig. 15.2: Mass distribution from BUU model calculation for $N_A = 120$ on $N_B = 120$ reaction at beam energies (a) 50 MeV/nucleon, (b) 75 MeV/nucleon, (c) 100 MeV/nucleon and (d) 150 MeV/nucleon. The average value of 5 mass units are shown. At each energy 1000 events are chosen. Only central collisions are considered here but even at $E_p = 50$ MeV/nucleon, nucleons in the peripheral region passes through and the largest fragment remaining is less than the sum of the masses of the two nuclei.

before disappearing. As the beam energy increases, the height of the second maximum decreases. At 75 MeV/nucleon the second maximum is still there but barely. At higher energy the multiplicity is monotonically decreasing and the fall becomes steeper with increasing beam energy. The evolution of this shape has long been known from CTM (Chapter 2). In CTM the natural variable is temperature T. In Fig. 15.3 we have shown the multiplicity distribution for a system of 192 particles at temperatures 6.5 MeV,

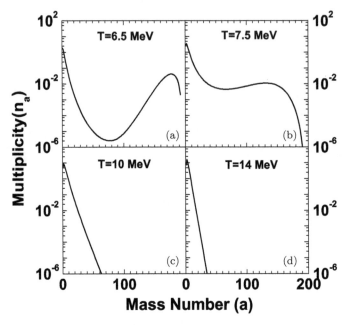

Fig. 15.3: Mass distribution from the CTM calculation for fragmentation of a system of mass $A_0 = 192$ at temperatures (a) 6.5 MeV, (b) 7.5 MeV, (c) 10 MeV and (d) 14 MeV.

7.5 MeV, 10 MeV and 14 MeV. The calculations with BUU and CTM are so different that the similarity in the evolution in shape of multiplicity is very striking. Indeed this correspondence provides great support for assumptions of statistical model from microscopic calculations.

This similarity in multiplicity distribution also suggests that nuclear forces will lead to first-order phase transition in heavy ion collisions at intermediate energy. In Fig. 15.2 the second maximum disappears at about 75 MeV/n beam energy. The disappearance of this maximum signals a first-order phase transition.

There are stronger theoretical signals that with just nuclear forces one sees the appearance of a first-order phase transition. Using our model of fluctuation we get multiplicity distribution. This can be used to check if bimodality (Chapter 7) emerges as the beam energy

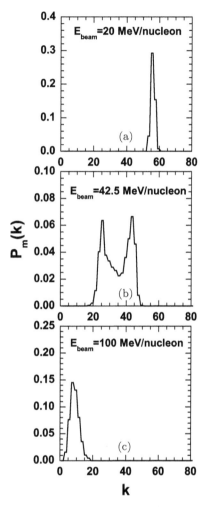

Fig. 15.4: Largest cluster probability distribution for $A_p = 40$ on $A_t = 40$ reaction at beam energies (a) 20 MeV/nucleon, (b) 42.5 MeV/nucleon and (c) 100 MeV/nucleon. The average value of 2 mass units are shown.

changes. This was done for 40 on 40 and 120 on 120 [115]. Since bimodality appears in a very narrow energy range one needs small steps in energy. Results for 40 on 40 are shown in Figs. 15.4 and 15.5. Figure 15.6 shows the results for 120 on 120.

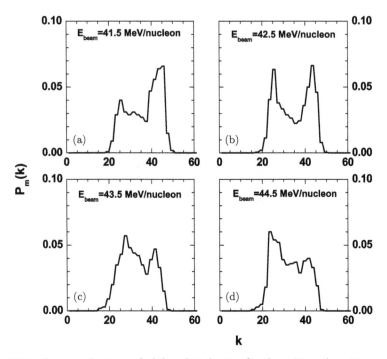

Fig. 15.5:   Largest cluster probability distribution for $A_p = 40$ on $A_t = 40$ reaction at beam energies (a) 41.5 MeV/nucleon, (b) 42.5 MeV/nucleon, (c) 43.5 MeV/nucleon and (d) 44.5 MeV/nucleon. The average value of 2 mass units are shown.

15.4   Discussion

We refer to one feature of this model that raised concerns and led to a lot of work [113,114] to propose alternative methods for calculations. This is related to dangers of crossing fermionic occupation limits in the model here. As mentioned, if Pauli blocking allows two tp's i and j to collide, then not only these two but also $\tilde{N} - 1$ tp's close to i and $\tilde{N} - 1$ tp's close to j move to represent that two actual nucleons scatter. Since the additional tp's are moved without verifying Pauli blocking there may be cases where one exceeds the occupation limits of fermions. For a discussion of this we refer to Appendix F.

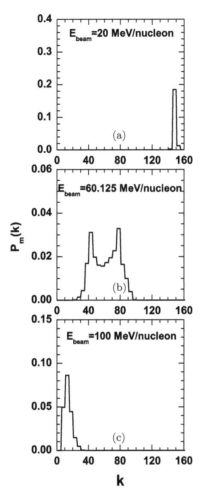

Fig. 15.6: Largest cluster probability distribution for $A_p = 120$ on $A_t = 120$ reaction at beam energies (a) 20 MeV/nucleon, (b) 60.125 MeV/nucleon and (c) 100 MeV/nucleon. The average value of 5 mass units are shown.

Chapter 16

Momentum Dependence

One primary goal of heavy ion collisions was to explore properties of nuclear systems at above and below normal nuclear density ($\rho_0 \approx 0.16\,\mathrm{fm}^{-3}$). A basic property is compressibility at normal density. This is normally quantified by K (see Chapter 11). From study of giant monopole vibration it was concluded that the value of K is about $215\,\mathrm{MeV}$ [116]. In the eighties this value began to be questioned. Experiments measuring flow angles [117] and transverse momenta [118] suggested that theories would give too small flow angles and too little transverse momenta with compressibility $K = 215\,\mathrm{MeV}$. A more careful parametrisation of nuclear EOS seems to have eliminated this discrepancy.

Danielewicz and Odyniec [119] proposed a transverse momentum analysis of experimental data. This is sensitive to the equation of state. Consider Nb on Nb. One defines a vector constructed from the transverse momenta of the detected particles

$$Q = \sum_{i=1}^{M} \omega_i P_{i\perp} \tag{16.1}$$

Here $\omega_i = 1$ if in the CM p_z of the detected particle is positive. If p_z is negative then ω_i is negative. One often uses rapidity $y = \frac{1}{2}\ln\frac{E+p_z}{E-p_z}$. Thus $\omega_i = +1$ if $y_i > 0$ and $\omega_i = -1$ if $y_i < 0$. Without the factor ω_i, Q is zero as implied by momentum conservation. With the factor ω_i the magnitude depends sensitively on the equation of state. This is the basic idea but some details are necessary. In a calculation the reaction plane is always known *a priori*. This is a plane formed by the impact parameter (x-direction) and the beam direction (z-direction). Symmetry rules out any component of Q in the y-direction. In experiments the reaction plane is not known *a priori* and $\vec{Q}$ determines the x-direction. In earlier work, particles near rapidity zero were excluded in the determination of $\vec{Q}$ but in later work (Doss) these are included. One can measure $p_x/A = \vec{Q} \cdot \vec{p}/QA$ as a function of rapidity where A is the number of nucleons in the particle whose momentum is being measured. Danielewicz and Odyniec point out because there are only a finite number of particles in an event, this gives an erroneous idea of the collective flow. Even in a random event generated by a Monte Carlo calculation that lacks a dynamic effect in the reaction plane, a significant amount of transverse momentum is found. By relating a particle to a construct in which it has been used, we probe a correlation of particle with itself. To remove the auto-correlation, one instead constructs $p_x/A = (\vec{Q} - \vec{p}) \cdot \vec{p}/|Q - p|A$. This then is the transverse momentum into a plane which is not a true reaction plane. The transverse momentum so measured will be less than the transverse momenta in the true reaction plane. In [118] the correction is estimated and experimental values of transverse momenta in the reaction plane are shown (see Fig. 16.1).

If one uses a purely density dependent mean field such as used in Chapter 11, one needs a high value of compressibility coefficient like $K = 380\,\mathrm{MeV}$ to fit the transverse momentum data. However it is well known that the mean field should have additional terms which depend on the momentum of the particle. The real part of the nuclear optical potential that a nucleon sees becomes less attractive as the nucleon energy increases. It goes through a zero and settles to a positive value. When this feature is built in the model one fits the transverse momentum data with K being close to $215\,\mathrm{MeV}$. Other experimental data, such as flow angles [117] that we have not touched, are also fitted quite well.

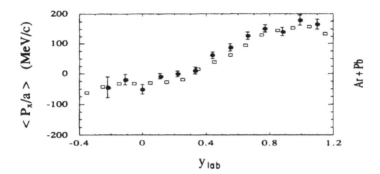

Fig. 16.1: Transverse momentum per nucleon as a function of rapidity in reactions of 800 MeV per projectile nucleon. Results of BUU simulations with the MDYI interaction (Eq. (16.8)) (open squares) are compared with the data.

Before we introduce an often used formula for a mean field which has a density dependent term as well as some momentum dependence, we write two well-known formulae from determinantal models. The total potential energy that a system has is given by

$$V = \frac{1}{2} \sum_{i,j} (v_{ijij} - v_{ijji}) \qquad (16.2)$$

and the potential energy that a nucleon in state i has is

$$U(i) = \sum_{j} (v_{ijij} - v_{ijji}) \qquad (16.3)$$

v_{ijkl} is a two-body matrix element:

$$v_{ijkl} = \iint \psi_i(\vec{r_1}) * \psi_j(\vec{r_2}) * v(r_{12}) \psi_k(\vec{r_1}) \psi_l(\vec{r_2}) d^3 r_1 d^3 r_2 \qquad (16.4)$$

The first term in Eqs. (16.2) and (16.3) is the direct term, the second term is the exchange term. The exchange term will lead to momentum dependence. i and j refer to orbitals. We can replace the sum over i, j by integration. Assuming that the system is in a volume

Ω we write

$$v_{ijji} = (1/\Omega^2) \iint e^{-i\vec{k}_i \cdot \vec{r_1}} e^{-i\vec{k}_j \cdot \vec{r_2}} v_0 \frac{e^{-\mu r_{12}}}{\mu r_{12}} e^{i\vec{k}_j \cdot \vec{r_1}} e^{i\vec{k}_i \cdot \vec{r_2}} d^3 r_1 d^3 r_2$$

(16.5)

The integral is easily done and after some algebra, the final answer is

$$v_{ijji} = \frac{4\pi v_0}{\Omega \mu^3} \frac{1}{1 + \frac{q^2}{\mu^2}}$$

(16.6)

where $q^2 = (\vec{k}_i - \vec{k}_j)^2 = (\vec{p}_i/\hbar - \vec{p}_j/\hbar)^2$. This is the momentum dependent part.

It is easily shown that the direct term $\sum_j v_{ijij}$ leads to $\frac{4\pi v_0}{\mu^3} \rho$ for $U(i)$. Thus the direct term can be absorbed in the ρ dependent term $A(\rho/\rho_0)$ that is always present in expressions for U.

We therefore write

$$V(\rho) = \frac{A}{2} \frac{\rho^2}{\rho_0} + \frac{B}{\sigma + 1} \frac{\rho^{\sigma+1}}{\rho_0^\sigma} + \frac{C}{\rho_0} \int d^3 p \int d^3 p' \frac{f(\vec{r}, \vec{p}) f(\vec{r}, \vec{p'})}{1 + [\frac{\vec{p} - \vec{p'}}{\Lambda}]^2}$$

(16.7)

This leads to a potential

$$U(\rho, \vec{p}) = A\frac{\rho}{\rho_0} + B\left(\frac{\rho}{\rho_0}\right)^\sigma + 2\frac{C}{\rho_0} \int d^3 p' \frac{f(\vec{r}, \vec{p'})}{1 + (\frac{\vec{p} - \vec{p'}}{\Lambda})^2}$$

(16.8)

Here $f(\vec{r}, \vec{p})$ is the phase space density. There are five constants in Eqs. (16.7) and (16.8). These are found by requiring that $E/A = -16\,\text{MeV}$, $\rho_0 = 0.16\,\text{fm}^{-3}$, $K = 215\,\text{MeV}$, $U(\rho_0, p = 0) = -75\,\text{MeV}$ and $U(\rho_0, p^2/2m = 300\,\text{MeV}) = 0$. The values are then $A = -110.44\,\text{MeV}$, $B = 140.9\,\text{MeV}$, $C = -64.95\,\text{MeV}$, $\sigma = 1.24$ and $\Lambda = 1.58 p_F^{(0)}$. This potential will be referred to as MDYI (momentum dependence from Yukawa interaction)[120].

An older version of momentum dependent interaction was used (and is still used) for heavy ion collisions [121]. It can be obtained from MDYI by replacing $\vec{p'}$ in the denominator of Eq. (16.7) by its average, $\langle \vec{p'} \rangle$ and is given the name V_{GBD}.

$$V_{GBD}(\rho(\vec{r})) = \frac{A}{2}\frac{\rho^2(\vec{r})}{\rho_0} + \frac{B}{\sigma+1}\frac{\rho^{\sigma+1}(\vec{r})}{\rho_0^\sigma}$$
$$+ \frac{C\rho(\vec{r})}{\rho_0}\int d^3p \frac{f(\vec{r},\vec{p})}{1+[\frac{\vec{p}-\langle\vec{p}\rangle}{\Lambda}]^2} \qquad (16.9)$$

The corresponding mean field is obtained by taking a functional derivative with respect to the single-particle occupation function: $U = \left(\frac{\delta V}{\delta f}\right)_p$. One then obtains

$$U_{GBD}(\rho(\vec{r},\vec{p})) = A\frac{\rho(\vec{r})}{\rho_0} + B\left(\frac{\rho(\vec{r})}{\rho_0}\right)^\sigma + \frac{C}{\rho_0}\int d^3p' \frac{f(\vec{r},\vec{p'})}{1+[\frac{\vec{p'}-\langle\vec{p'}\rangle}{\Lambda}]^2}$$
$$+ \frac{C}{\rho_0}\frac{\rho}{1+[\frac{\vec{p}-\langle\vec{p}\rangle}{\Lambda}]^2} \qquad (16.10)$$

The parameters chosen were [123] $A = -144\,\mathrm{MeV}$, $B = 203.3\,\mathrm{MeV}$, $C = -75\,\mathrm{MeV}$, $\sigma = 7/6$ and $\Lambda = 1.5\,p_F^{(0)}$.

The $U(\rho,p)$ calculated using MDYI Eq. (16.8) can be compared with other calculations. Wiringa [122] has calculated the single-particle potential in nuclear matter using several realistic Hamiltonians for densities ranging from 0.1 to $0.5\,\mathrm{fm}^{-3}$. These Hamiltonians include nucleon–nucleon potentials fit to scattering data and three-nucleon potentials fit to binding energies of few-body nuclei and saturation properties of nuclear matter. The results of these microscopic calculations were parametrised and the values of the parameters are given in Table I of [122]. This table is used to compare Wiringa's calculations with $U(\rho,p)$ obtained from MDYI, Eq. (16.8). Comparisons with results of UV14 + TNI (Fig. 16.2) and UV14 + UVII (Fig. 16.3) show that $U(\rho,p)$ obtained from MDYI is quite reasonable. One can also compare with U_{GBD} Eq. (16.10). The fit is not as good but acceptable.

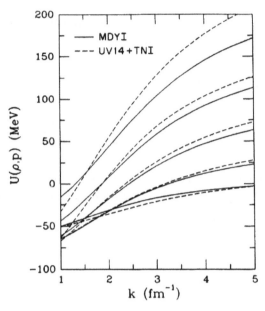

Fig. 16.2: A comparison of the single-particle potential from MDYI with the microscopic calculations of Wiringa [122] using the UV14 + TNI interaction. The abscissa shows wave numbers. Starting from the bottom at right, the different curves are for densities of 0.1, 0.2, 0.3, 0.4 and 0.5 fm^{-3}.

To use MDYI interaction, readers may find the following equations useful. These are taken from [120].

$$\int_0^{p_F} \int_0^{p_F} \frac{d^3p\, d^3p'}{1 + \left[\frac{\vec{p}-\vec{p}'}{\Lambda}\right]^2} = \frac{32\pi^2 p_F^4 \Lambda^2}{3}\left[\frac{3}{8} - \frac{\Lambda}{2p_F}\tan^{-1}\left(\frac{2p_F}{\Lambda}\right) - \frac{\Lambda^2}{16p_F^2}\right.$$

$$\left. + \left\{\frac{3\Lambda^2}{16p_F^2} + \frac{\Lambda^4}{64p_F^4}\right\}\ln\left\{1 + \frac{4p_F^2}{\Lambda^2}\right\}\right] \quad (16.11)$$

$$\int_0^{p_F} \frac{d^3p'}{1 + \left[\frac{\vec{p}-\vec{p}'}{\Lambda}\right]^2} = \pi\Lambda^3\left[\frac{p_F^2 + \Lambda^2 - p^2}{2p\Lambda}\ln\left\{\frac{(p+p_F)^2 + \Lambda^2}{(p-p_F)^2 + \Lambda^2}\right\} + \frac{2p_F}{\Lambda}\right.$$

$$\left. + 2\left\{\tan^{-1}\left(\frac{p+p_F}{\Lambda}\right) - \tan^{-1}\left(\frac{p-p_F}{\Lambda}\right)\right\}\right] \quad (16.12)$$

We now consider a different experiment which also tests the equation of state. This is measurement of flow angles. The experiment

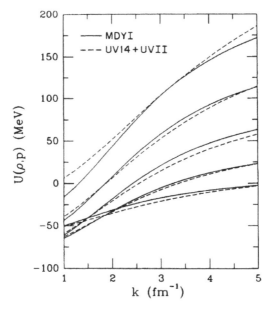

Fig. 16.3:   A comparison of the single-particle potential from MDYI with the microscopic calculations of Wiringa [122] using the UV14+UVII interaction. The abscissa shows wave numbers. Starting from the bottom at right, the different curves are for densities of 0.1, 0.2, 0.3, 0.4 and 0.5 fm^{-3}.

uses 4π detectors. In the experiment momentum and mass of each charged particle are measured. The number of charged particles depends upon the impact parameter. By restricting the observation to high multiplicity events, one can select on nearly central collisions. For each event one measures the six elements of the symmetric matrix

$$F_{ij} = \sum_{\nu} \frac{p_i(\nu)p_j(\nu)}{2m(\nu)} \qquad (16.13)$$

Here i, j are the three Cartesian coordinates, $m(\nu)$ the particle mass. Diagonalise the matrix. Let θ be the angle (with respect to the original beam direction) of the eigenvector corresponding to the highest eigenvalue; θ is called the flow angle. The data, $dN/d(\cos\theta)$ can be plotted as a histogram. In calculations the maximum in the histogram appears at an angle θ which depends upon the equation of state (EOS). For a soft EOS without any momentum dependence the

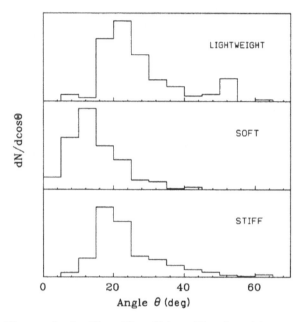

Fig. 16.4: Flow angle for Nb + Nb collision. The three histograms show the results from the lightweight (upper panel), the soft (middle panel) and the stiff (lower panel) models.

maximum appears around 15 degrees (too small compared to what is seen in experiments on Nb on Nb at $400\,\mathrm{MeV/n}$). The maximum moves to a larger angle with a hard EOS without momentum dependence but not enough. Inclusion of momentum dependence in a soft EOS reproduces the experimental flow angle quite well. The results of the three theoretical calculations, taken from [121], are shown in Fig. 16.4. The result with momentum dependence is labelled as lightweight in the figure. The momentum dependent part of the interaction can be written as $\frac{\alpha p_F^2(0)}{2m}$ at $\rho_0, p_F(0)$. With our interaction α is about 0.3. The effect of this can be absorbed by defining an effective mass $\frac{m^*}{m} \approx 0.7$ where $\frac{p_F^2(0)}{2m} + \frac{\alpha p_F^2(0)}{2m} = \frac{p_F^2(0)}{2m^*}$. As $\frac{m^*}{m}$ is less than 1 it is called a lightweight EOS.

BUU with density and momentum dependent interactions was carried out with test particle method several times [120,121,123,124]. In recent calculations typically the cell size was $1.5\,\mathrm{fm}^3$ and number of test particles 150 per nucleon. What is very interesting is that with MDYI and GBD, significant transverse momenta are generated even by pure Vlasov propagation (no hard collisions). There are more quantities to take care of with momentum dependence in. For example, when updating velocities one has to use $\frac{d\vec{r_i}}{dt} = \frac{\vec{p_i}}{m} + \vec{\nabla}_p U$. One is required to store more quantities to advance in time. It is also obvious that hard collisions would not conserve energy exactly. This and even effects of non-conservation of angular momentum have been estimated [124]. It is beyond the limitations of this book to address these questions and we refer to the above mentioned publication.

Another test to examine the nuclear EOS was the measurement of pion production at intermediate energy. In statistical model, for example, the pion cross-sections have been too high. This is so with a soft EOS without momentum dependance and improves very little with a hard EOS. In fact, the overestimation of pion multiplicity was the basis for the concept that the number of pions might be used to deduce the nuclear EOS [125]. This was based on the assumption that (a) the pion multiplicity is determined at the point of maximum compression and (b) the cascade model is deficient only in their lack of mean-field effect. Both assumptions appear to be wrong. We use the frozen Delta model [87] but here is an important point. The Delta mass distribution has its experimentally observed shape which is not of the fixed width Breit–Wigner type. A parametrisation is given in [126]. If this is used the agreement with experiment becomes quite good. To see why the agreement improves we refer to Gale [127] but here, for brevity, we will just write the steps that are needed to calculate the M_Δ.

$$f(M) = \frac{\frac{\Gamma^2(q)}{4}}{\left[(M - M_0)^2 + \frac{\Gamma^2(q)}{4}\right]} \tag{16.14}$$

with $M_0 = 1232\,\text{MeV}$. The width $\Gamma(q)$ is given by the phenomenological equation [128]

$$\Gamma(q) = \frac{0.47}{\left[1 + 0.6\left(\frac{q}{M_\pi}\right)^2\right]} \frac{q^3}{M_\pi^2} \tag{16.15}$$

where the CM momentum q is related to the quantity M as

$$M = (M_N^2 + q^2)^{1/2} + (M_\pi^2 + q^2)^{1/2} \qquad (16.16)$$

with the rest mass of pion and nucleon, M_π and M_N.

When two nucleons collide inelastically with CM energy $\sqrt{s}$ the mass M of the Delta is chosen from

$$x = \frac{\int_{M_N+M_\pi}^{M} f(M)dM}{\int_{M_N+M_\pi}^{\sqrt{s}-M_N} f(M)dM} \qquad (16.17)$$

Momentum dependence of the mean field plays a minor role here. It slightly lowers the Delta production.

Aichelin *et al.* have also investigated the effect of momentum dependent interactions [129]. They used QMD [130] (to be briefly discussed in Chapter 17) and found that transverse momenta are strongly augmented by momentum dependence [129].

Chapter 17

Quantum Molecular Dynamics Model

The nuclei we will deal with in heavy ion collisions have of the order of 200 nucleons. The time evolution of a system of these many particles, if they follow classical physics rules, can be traced out using what is called molecular dynamics. Vicentini *et al.* [131] did make such a study which brought out well-known features of co-existence, vaporisation etc. which are also expected in quantal systems like nuclei. While a fully quantal many-body treatment is of course not practical, in what is now labelled as "quantum molecular dynamics" one can compute in a model transverse momenta, collective flow (described in Chapter 16), multifragmentation (although many applications for this aspect are not available). In the model, effects of Pauli blocking are included when nucleon–nucleon collisions are considered. We describe the main features of the model.

The most important feature of the model is that nucleons are represented by wave packets with a width in both coordinate and momentum space. Instead of wavefunctions one uses Wigner transforms $f(\vec{r}, \vec{p})$ for each nucleon. At time $t = 0$ a nucleon is represented by

$$f(\vec{r}_i, \vec{p}_i) = \left(\frac{\alpha}{\sqrt{\pi}}\right)^3 \exp\{-\alpha^2(\vec{r}_i - \vec{r}_{0i})^2\}$$

$$\times \left(\frac{1}{\hbar\alpha\sqrt{\pi}}\right)^3 \exp\{-(\vec{p}_i - \vec{p}_{0i})^2/\hbar^2\alpha^2\}$$

$$= \left(\frac{1}{\hbar\pi}\right)^3 \exp\{-\alpha^2(\vec{r}_i - \vec{r}_{0i})^2\} \exp\{-(\vec{p}_i - \vec{p}_{0i})^2/\hbar^2\alpha^2\}$$

$$(17.1)$$

The crucial parameter here is α^2 which controls the width in configuration space. The width is left unaltered throughout the calculation. The centroid $\vec{r}_{0i}$ is a function of time and signifies the movement of the nucleon i in configuration space. The zero time choice for $\vec{p}_{0i}$ needs some discussion. But once this is chosen, any subsequent change is dictated by the model. Usually it changes slowly except when hard collisions occur (just like in BUU). The width of the momentum wave packet does not change, just like the width of the wave packet in configuration space. The choice of α is dictated by the requirement that at initialisation the nuclear density should be reasonably smooth within a sphere of radius $R = 1.18A^{1/3}$ where A is the mass number of the nucleus. The value of α^2 is chosen to be about $0.5\,\text{fm}^{-2}$. With this choice the square root of the mean square radius of a nucleon is $1.8\,\text{fm}$. By ensuring that the distance between the centres of any pair of wave packets in r-space is not less than $1.5\,\text{fm}$ and this choice of α^2, a smooth density can be achieved. Next is the question of the choice of the initial $\vec{p}_{0i}$. The choice has changed over the years. Originally [130] at point $\vec{r}_{0i}$ the density was computed by adding contributions from all the wave packets in r-space and a value of $p_{F01}(\vec{r}_{0i})$ obtained by local density approximation. The initial $\vec{p}_{0i}$ was then Monte Carloed from a sphere of radius of this p_F. We will have to wait a little to introduce the more recent procedure of picking the initial value of $\vec{p}_{0i}$ [132].

When the model was introduced, the interactions were 2-body delta functions $t_1\delta(\vec{r_1} - \vec{r_2})$ and 3-body delta functions $t_2\delta(\vec{r_1} - \vec{r_2})\delta(\vec{r_1} - \vec{r_3})$. Later Coulomb interaction, Yukawa interaction and a momentum dependent interaction were added. Here we will limit to the delta function interactions. The centroids in r-space play special roles in QMD. The two-body interaction $t_1(\vec{r_1} - \vec{r_2})$ is not just the interaction between the two centroids directly as the centroids are almost always apart (initially they are always apart; that is how they were created). This is how the wave packet with centroid at $\vec{r_{0i}}$ is influenced by the wave packet with centroid at $\vec{r_{0j}}$ by the 2-body delta force. The tail of the packet with centroid at $\vec{r_{0j}}$ will be non-zero at $\vec{r_{0i}}$. Thus $U(\vec{r_{0i}})$ due to centroid at $\vec{r_{0j}}$ is $C_1\exp\{-\alpha^2(\vec{r_{0i}} - \vec{r_{0j}})^2\}$ where C_1 is in MeV and has absorbed all factors. It follows of course that the influence of the centroid at $\vec{r_{0i}}$ on $U(\vec{r_{0j}})$ is identical. The contribution of three-body delta functions on $U(\vec{r_{0i}})$ is $C_2[\sum_{k\neq i}\exp(-\alpha^2(\vec{r_{0i}} - \vec{r_{0k}})^2)]^2$. C_2 is in MeV and has absorbed all factors like t_2, $\hbar$ etc. In squaring the sum above there is some overcounting but that has negligible effect.

We mentioned how at initialisation the centroids of the wave packets in r-space, their width and the width of the momentum packets are chosen. To complete the story, one needs to know how the centroid of the momentum packet $\vec{p_{0i}}$ is chosen. The local Fermi momentum is determined by the relation $\sqrt{-2mU(\vec{r_{0i}})}$. Finally the momentum of $\vec{p_{0i}}$ is chosen randomly between 0 and the local Fermi momentum. One then rejects all random numbers which yield two particles closer in phase space than $(\vec{r_{0i}} - \vec{r_{0j}})^2 \times (\vec{p_{0i}} - \vec{p_{0j}})^2 = d_{min}$ where d_{min} is some pre-set value. Typically only 1 out of 50,000 initialisations is accepted under the present criteria. The computational time required for the initialisation is short compared to the time needed for propagation [132].

Let us consider collision between nuclei A and B. We will summarise the propagation with mean field but will give some details about hard scattering. Normally the centroids of the wave packets will be boosted towards each other and move smoothly with velocity $\vec{v}(0)$. The centroids $\vec{r_{0i}}, \vec{p_{0i}}$ will continue to change by following equations:

$$\vec{r}_{0i}(t + \delta t) = \vec{r}_{0i}(t) + \frac{\vec{p}_{i0}(t + 0.5\delta t)}{m}\delta t$$

$$\vec{p}_{0i}(t + 0.5\delta t) = \vec{p}_{0i}(t - 0.5\delta t) - \vec{\nabla}_r U(\vec{r}_{0i}(t))\delta t \qquad (17.2)$$

Hard collisions between two centroids in r-space and p-space follow exactly the same procedure of n–n hard collisions as detailed in Appendix B or Appendix C. In hard scattering the centroids will change suddenly from $\vec{r}_{0i}, \vec{p}_{0i}$ to $\vec{r}_{0i}', \vec{p}_{0i}'$ and $\vec{r_{0j}}, \vec{p_{0j}}$ to $\vec{r_{0j}'}, \vec{p_{0j}'}$ if the transition is not Pauli blocked. Although prescriptions to check Pauli blocking have evolved, one procedure is this. For simplicity assume that each nucleon occupies a sphere in r-space and p-space. One determines which fractions P_1 and P_2 of the final phase spaces for each of the scattering partners are already occupied by other nucleons. The probability of blocking scattering is $1 - [1 - min(P_1, 1)][1 - min(P_2, 1)]$. This is Monte Carloed. If the scattering is blocked we revert back to original centroids. For identifying clusters, in QMD approach minimum spanning tree (MST) method, simulated annealing clusterisation algorithm (SACA) method [133] and fragment recognition in general application (FRIGA) model [134] are commonly used.

There has been a great deal of work with a model known as AMD (antisymmetrised molecular dynamics) which, in few aspects, is similar to QMD, and in many aspects, very different. The interacting system is represented by antisymmetrised many-body wave functions consisting of single-particle states with spin and isospin which are localised in phase space. We give a bare outline and refer to original papers as this is outside the scope of this monograph. The single-particle orbitals are complex (to allow wave packets with net momenta; these are not semi-classical Wigner transforms but quantum mechanical wave packets) and each orbital carries spin and isospin. The wave packets are Gaussian and the width parameter is treated as time-independent. The time developments of the centres of the Gaussian wave packets are determined by two processes. One is the time development determined by variational principle, which takes the place of mean-field propagation. The second process that determines the time development of the system is the stochastic collision process due to the residual interaction. This is done in a similar way to QMD however it is clearly not straightforward and some steps need to be carried out to determine what should be used as position and momentum coordinates of nucleons [135–137]. One big advantage of AMD is that calculations with finite range effective interaction is as feasible as calculations with zero range effective interaction and therefore the momentum dependence of the mean field, which is essential for the study of the EOS is automatically included in the calculation. A different approach to AMD can be found in [138].

Chapter 18

Epilogue

The last forty years were an exciting period in intermediate energy heavy ion collisions. From late seventies to early nineties many pioneering experiments were launched at Bevalac at the Lawrence Berkeley Lab. The topic that attracted a great deal of attention was the determination of compressibility coefficient K of nuclear matter. Is it 215 MeV or is it closer to 300 MeV? The question inspired theorists to formulate microscopic models with which they could compute several experimental data which seemed to relate to EOS of nuclear matter which involves, among other aspects, compressibility. The development of the microscopic model was a remarkable breakthrough by simultaneously including both mean field and pure n–n collisions. This is now considered part of standard nuclear physics. This monograph described how an improved microscopic model re-established K to be about 215 MeV.

Along with microscopic models, there were developments in macroscopic models. There were various versions but the standard theme behind most of these is that in intermediate energy heavy ion collisions there are so many open channels that phase space decides what will come out. For the few Bevalac experiments that were tried, the idea worked moderately well but the thermodynamics based models really took off when experiments at lower beam energy began to be done in many labs. In particular Copenhagen model got absolutely great fits with many data. These are great examples of thermodynamics at work. These calculations are lengthy but in 1995 a mathematical algorithm was published which made the canonical thermodynamic model well within everybody's reach. A huge amount of interesting data have been fitted with this model. This should become part of standard nuclear theory and statistical mechanics.

Another exciting possibility was observing phase transition in intermediate energy heavy ion collisions. The nucleus is thought of as a liquid drop so for a large nuclear system (such as can be formed in heavy ion collisions) one might discover signatures of phase transition. There were many different ideas: continuous phase transitions, first-order phase transition etc. Textbook examples like the lattice gas model have been applied to nuclear physics case. Heavy ion collision provides a great many examples of interesting statistical physics. Experimentally caloric curve resembling that of first-order phase transition was found.

The microscopic models and thermodynamic models generally explored different observables and one might wonder if one can go from one to the other. They can be combined to get interesting results.

A modified microscopic model with fluctuations produced multiplicity distributions similar to that resulting from thermodynamic models. The calculations are so different that the similarity of the final results are quite amazing.

We will finish this epilogue with one more interesting thing. Let us refer to isoscaling (Chapter 4). Let us denote by R_{21} the ratio of the yields of the same isotope (N, Z) from two different reactions 1 and 2, $R_{2,1} = Y_2(N, Z)/Y_1(N, Z)$. It is found that in many cases, the ratio is fitted extremely well by Eq. (4.1): $R_{21} = C \exp(\alpha N + \beta Z)$. As explained in Chapter 4, the result follows from grand canonical model exactly. Such accurate fitting may be considered a triumph of the grand canonical assumption. We like to state even when this parametrisation fails, it is understandable. For grand canonical model to be valid one needs a system to be in diffusive contact with a reservoir which is much bigger than the system. We may regard $Y(N, Z)$ to be the system and the fireball from which it comes to be the reservoir. In some cases the system is a lot smaller than the reservoir, then isoscaling emerges. In some cases the system is almost as large as the reservoir, then isoscaling would not work. The examples of both situations are shown in Fig. 4.2.

To summarise, intermediate energy heavy ion collision is full of many interesting observables and many interesting models. We expect this to continue.

Appendix A

Volume of Participant

Consider the collision between two unequal ions at an impact parameter b (Fig. A.1) where ion B is smaller than ion A. Collision occurs when b is between 0 and $R_A + R_B$. Break up the collision into collisions between thin discs of thickness Δy at distance y from the centres. The maximum value of y (denoted by y_m) is R_B if $b < (R_A^2 - R_B^2)^{1/2}$ else it is determined by the equation

$$(R_B^2 - y_m^2)^{1/2} + (R_A^2 - y_m^2)^{1/2} - b = 0 \qquad \text{(A.1)}$$

Let $R_A(y) = (R_A^2 - y^2)^{1/2}$ and $R_B(y) = (R_B^2 - y^2)^{1/2}$. Then the volume $V_A(b)$ that the colliding part of B gouges out of A is givenby

$$V_A(b) = 4 \int_0^{y_m} dy \int_{x_1}^{x_2} (R_A^2(y) - x^2)^{1/2} dx \qquad \text{(A.2)}$$

The lower limit x_1 is $b - R_B(y)$ and the upper limit is either $R_A(y)$ or $b + R_B(y)$, whichever is smaller. Similarly

$$V_B(b) = 4 \int_0^{y_m} dy \int_{x_1}^{x_2} (R_B^2(y) - (b - x)^2)^{1/2} dx \qquad \text{(A.3)}$$

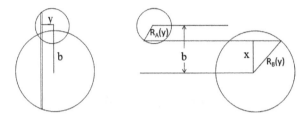

Fig. A.1: Geometrical analysis of abrasion stage.

Appendix B

Cascade without Pion

Computation of n–n collision is a crucial part of a transport code. There are many such codes available. For completeness we describe one which was introduced in the eighties [90] and is still being used. We use cross-section in mb, $\sqrt{s}$ in GeV, momentum in GeV/c and $c = 1$.

In the eighties many experiments were done in the 400 to 800 MeV beam energy range with emphasis on near central collisions. Measurement of pion production was one of the goals. These pions would result from n–n collisions. After Bevalac was switched off, experiments in the higher end of intermediate energy became rarer. Many more experiments are at lower energy and we expect to see more and more use of transport codes in this lower energy range. In such cases one can ignore the inelastic channel in n–n collision and the programming becomes simpler. This appendix deals with this situation. The more general case which includes the inelastic channel is given in Appendix C.

For elastic channels: $n + n \rightarrow n + n$, if $\sqrt{s} < 1.8993$ use

$$\sigma^{el}(\sqrt{s}) = 55 \tag{B.1}$$

If $\sqrt{s} > 1.8993$ then

$$\sigma^{el} = \frac{35}{1 + 100(\sqrt{s} - 1.8993)} + 20 \tag{B.2}$$

The differential cross-section for all elastic scattering is taken to be

$$d\sigma/dt = ae^{bt} \tag{B.3}$$

Here $t = -2p^2(1 - \cos\theta)$ and is the negative of the square of momentum transfer in the CM of the scattering particles; θ_s goes from 0 to $\pi/2$ and t from $t_0 = -2p^2$ to 0. The parameter $b(\sqrt{s})$ is chosen to be

$$b(\sqrt{s}) = \frac{6[3.65(\sqrt{s} - 1.866)]^6}{1 + [3.65(\sqrt{s} - 1.866)]^6} \tag{B.4}$$

The value of t_1 is chosen from

$$\frac{\int_{t_0}^{t_1} ae^{bt}dt}{\int_{t_0}^{0} ae^{bt}dt} = x_1 \tag{B.5}$$

where x_1 is a random number. The value of θ_s is chosen from $\cos\theta_s = 1 - t_1/t_0$; ϕ_s is chosen randomly. In the collisions considered, momentum and energy are conserved but not angular momentum. The cumulative effect is small.

We now need to decide if in a given time interval $-\delta t/2$ to $\delta t/2$ two nucleons will collide. We have $\sigma^t(\sqrt{s}) \leq \sigma^t_{max} = 55$ mb. This gives $b_{max} = 1.32$ fm. Thus if the two particles are apart by more than $\sqrt{(1.32)^2 + c^2(\delta t)^2}$ they cannot collide in time δt. If the distance is less we check if the two particles pass the point of closest approach and the distance of closest approach is less than 1.32 fm.

The CM energy is $\sqrt{s} = \sqrt{(E_1 + E_2)^2 - (\vec{p_1} + \vec{p_2})^2}$. The velocity of the CM of the particles is $\vec{\beta} = (\vec{p_1} + \vec{p_2})/(E_1 + E_2)$. The momentum of particle 1 in the CM frame is

$$\vec{p} = \gamma\left(\frac{\vec{p_1}\cdot\vec{\beta}}{\beta} - \beta E_1\right)\frac{\vec{\beta}}{\beta} + \left(\vec{p_1} - \frac{\vec{p_1}\cdot\vec{\beta}}{\beta}\frac{\vec{\beta}}{\beta}\right) \tag{B.6}$$

The momentum of the second particle is $-p$. The distance $\vec{\Delta r}$ in the CM frame is $(\gamma - 1)[(\vec{r_1} - \vec{r_2})\cdot\vec{\beta}/\beta]\vec{\beta}/\beta + (\vec{r_1} - \vec{r_2})$. In the time interval $-\delta t/2$ to $+\delta t/2$ the two particles become candidates for collision if $|\frac{\vec{\Delta r} \cdot \vec{p}}{p}| < (\frac{p}{\sqrt{p^2+m_1^2}} + \frac{p}{\sqrt{p^2+m_2^2}})\delta t/2$ and $b = \sqrt{(\vec{\Delta r})^2 - (\vec{\Delta r} \cdot \vec{p}/p)^2}$ is less than 1.32 fm. The following sequence of operations is now performed:

1. The elastic cross-section $\sigma_{nn}^e(\sqrt{s})$ is computed from Eqs. (B.1) and (B.2). Generate a random number h. If $h < \sigma_{nn}^e(\sqrt{s})/55$, elastic scattering occurs and the angle is chosen according to Eq. (B.5). Then we branch to step 3.

2. If $h > \sigma_{nn}^e(\sqrt{s})/55$, no scattering occurred, $\vec{p}$ did not change. Skip step 3 and go to step 4.

3. The vector $\vec{p}$ has changed to $\vec{p}'$. Let us suppose that the original vector $\vec{p}$ was in the z' direction with original angles $\theta = \cos^{-1} p_z/p$ and $\phi = \tan^{-1} p_y/p_x$. After scattering z' will be further deviated by scattering angles θ_s and ϕ_s: the values of θ_s and ϕ_s are already known.

We can regard z' as the z-axis of a rotated frame where the first rotation is ϕ about the z-axis to obtain x_1, y_1-axes and the second rotation is θ about y_1-axis. If we refer to rotation about fixed axes then the order of rotation is reversed: thus $z' = \exp(-i\phi l_z)\exp(-i\theta l_y)z$. The final direction is ϕ_s about z' and then θ_s. Thus the final direction $\vec{p}'$ is given by

$$\vec{p}'/p' = \exp(-i\phi l_z)\exp(-i\theta l_y)\exp(-i\phi_s l_z)\exp(-i\theta_s l_y)z \qquad (B.7)$$

After some algebra one finds

$$p_x' = |p'|[\cos\phi(\cos\theta\sin\theta_s\cos\phi_s + \cos\theta_s\sin\theta) - \sin\theta_s\sin\phi\sin\phi_s]$$

$$p_y' = |p'|[\sin\phi(\cos\theta\sin\theta_s\cos\phi_s + \cos\theta_s\sin\theta) + \sin\theta_s\cos\phi\sin\phi_s]$$

$$p_z' = |p'|[\cos\theta\cos\theta_s - \sin\theta\sin\theta_s\cos\phi_s]$$

4. We now move back from the CM of the two particles to the original frame. If we came from step 2, the momenta of the particles did not change. If we came from step 3, the momenta changed. Once all the collisions are checked, the particles propagate with $\vec{r_i}(t+\Delta t) = \vec{r_i} + \Delta t(\vec{p_i}/\sqrt{p_i^2 + m_i^2})$.

Appendix C

Cascade with Pion

Computation of n–n collision is a crucial part of transport code. There are many such codes available. For completeness we describe one which was introduced in the eighties [95] and still being used. We use cross-section in mb, $\sqrt{s}$ in GeV, momentum in GeV/c and $c = 1$.

For elastic channels: $n + n \rightarrow n + n; n + \Delta \rightarrow n + \Delta; \Delta + \Delta \rightarrow \Delta + \Delta$, if $\sqrt{s} < 1.8993$ use

$$\sigma^{el}(\sqrt{s}) = 55 \tag{C.1}$$

If $\sqrt{s} > 1.8993$ then

$$\sigma^{el} = \frac{35}{1 + 100(\sqrt{s} - 1.8993)} + 20 \tag{C.2}$$

Cross-section for inelastic channel $n + n \rightarrow n + \Delta$ is 0 if $\sqrt{s} < 2.015$. For $\sqrt{s} > 2.015$ it is

$$\sigma^{in}_{nn \rightarrow n\Delta} = \frac{20(\sqrt{s} - 2.015)^2}{0.015 + (\sqrt{s} - 2.015)^2} \tag{C.3}$$

Cross-section $\sigma_{n\Delta \rightarrow nn}$ is given by detailed balance

$$\sigma_{n\Delta \rightarrow nn} = (p_f^2/p_i^2)\frac{1}{8}\sigma_{nn \rightarrow n\Delta} \tag{C.4}$$

where p_f is the momentum in the final channel. The factor 8 accounts for spin–isospin and identical nature of particles in the final channel. All calculations are performed in the CM of the colliding particles.

The differential cross-section for all elastic scattering is taken to be

$$d\sigma/dt = ae^{bt} \qquad (C.5)$$

Here $t = -2p^2(1 - \cos\theta)$ and is the negative of the square of momentum transfer in the CM of the scattering particles; θ_s goes from 0 to $\pi/2$ and t from $t_0 = -2p^2$ to 0. The parameter $b(\sqrt{s})$ is chosen to be

$$b(\sqrt{s}) = \frac{6[3.65(\sqrt{s} - 1.866)]^6}{1 + [3.65(\sqrt{s} - 1.866)]^6} \qquad (C.6)$$

The value of t_1 is chosen from

$$\frac{\int_{t_0}^{t_1} ae^{bt}dt}{\int_{t_0}^{0} ae^{bt}dt} = x_1 \qquad (C.7)$$

where x_1 is a random number. The value of θ_s is chosen from $\cos\theta_s = 1 - t_1/t_0$; ϕ_s is chosen randomly. In the collisions considered, momentum and energy are conserved but not angular momentum. The cumulative effect is small. If isotropy is assumed in inelastic scattering the angle of scattering can be chosen by $\cos\theta_s = 1 - 2x$.

We now need to decide if in a given time interval $-\delta t/2$ to $\delta t/2$ two nucleons will collide. We have $\sigma^t(\sqrt{s}) \leq \sigma^t_{max} = 55\,\mathrm{mb}$. This gives $b_{max} = 1.32\,\mathrm{fm}$. Thus if the two particles are apart by more than $\sqrt{(1.32)^2 + c^2(\delta t)^2}$ they cannot collide in time δt. If the distance is less we check if the two particles pass the point of closest approach and the distance of closest approach is less than 1.32 fm.

The CM energy is $\sqrt{s} = \sqrt{(E_1 + E_2)^2 - (\vec{p_1} + \vec{p_2})^2}$. The velocity of the CM of the particles is $\vec{\beta} = (\vec{p_1} + \vec{p_2})/(E_1 + E_2)$. The momentum of particle 1 in the CM frame is

$$\vec{p} = \gamma\left(\frac{\vec{p_1} \cdot \vec{\beta}}{\beta} - \beta E_1\right)\frac{\vec{\beta}}{\beta} + \left(\vec{p_1} - \frac{\vec{p_1} \cdot \vec{\beta}}{\beta}\frac{\vec{\beta}}{\beta}\right) \qquad (C.8)$$

The momentum of the second particle is $-p$. The distance $\vec{\Delta r}$ in the CM frame is $(\gamma - 1)[(\vec{r_1} - \vec{r_2}) \cdot \vec{\beta}/\beta]\vec{\beta}/\beta + (\vec{r_1} - \vec{r_2})$. In the time interval

$-\delta t/2$ to $+\delta t/2$ the two particles become candidates for collision if $\left|\frac{\vec{\Delta r}\cdot\vec{p}}{p}\right| < \left(\frac{p}{\sqrt{p^2+m_1^2}} + \frac{p}{\sqrt{p^2+m_2^2}}\right)\delta t/2$ and $b = \sqrt{(\vec{\Delta r})^2 - (\vec{\Delta r}\cdot\vec{p}/p)^2}$ is less than 1.32 fm. The following sequence of operations is now performed:

1. The elastic cross-section $\sigma_{nn}^e(\sqrt{s})$ is computed from Eqs. (C.1) and (C.2). Generate a random number h. If $h < \sigma_{nn}^e(\sqrt{s})/55$, elastic scattering occurs and the angle is chosen from Eq. (C.7). Then we branch to step 5. If $h > \sigma_{nn}^e(\sqrt{s})/55$, go to step 2 to examine the possibility of inelastic scattering.

2. If $\sqrt{s} < 2.015$ both are nucleons ($m_n = 0.938$) but there is not enough energy to produce a Δ. We branch to step 4.

If m_1 and m_2 are both greater than 0.938 (both are Δ's) there is no possibility of inelastic scattering and we branch to step 4.

Compute the inelastic cross-section $\sigma_{nn\to n\Delta}^{in}$ Eq. (C.3). If one of the colliding particles is a nucleon and the other one is a Δ, we branch to step 3. Otherwise compare h with $(\sigma_{nn\to n\Delta}^{in} + \sigma_{nn}^{el})/55$. If h is greater, branch to step 4. If h is less, then Δ production will occur. One has to choose the mass m_Δ. Different prescriptions have been used and the number of pions predicted depends on them [127]. We will have more to say about this later. A simple one is $m_\Delta = 1.077 + 0.75(\sqrt{s}-2.015)$ for $2.015 < \sqrt{s} < 2.2203$; $m_\Delta = 1.231$ for $\sqrt{s} > 2.2203$ [90]. Having determined m_Δ the magnitude of p_f is fixed by energy conservation. The angles are determined by assumed isotropy of scattering. Branch to step 5.

3. Since $\sigma_{nn\to n\Delta}^{in}$ has been computed, obtain $\sigma_{n\Delta\to nn}^{in}$ from Eq. (C.4). Compare h with $(\sigma_{n\Delta\to nn}^{in})/55$. If h is greater, we go to step 4. If h is less, then we have Δ absorption. The final momentum is obtained from energy conservation and isotropy. Now branch to step 5.

A much better prescription for pion production is given in Chapter 16 Eqs. (16.14) to (16.17). We will not repeat the argument here.

4. Since no scattering occurred, $\vec{p}$ did not change. We skip step 5 and branch to step 6.

5. The vector $\vec{p}$ has changed to $\vec{p'}$. Let us suppose that the original vector $\vec{p}$ was in the z' direction with original angles $\theta = \cos^{-1} p_z/p$ and $\phi = \tan^{-1} p_y/p_x$. After scattering z' will be further deviated by scattering angles θ_s and ϕ_s: the values of θ_s and ϕ_s are already known.

We can regard z' as the z-axis of a rotated frame where the first rotation is ϕ about the z-axis to obtain x_1, y_1-axes and the second rotation is θ about y_1-axis. If we refer to rotation about fixed axes then the order of rotation is reversed: thus $z' = \exp(-i\phi l_z)\exp(-i\theta l_y)z$. The final direction is ϕ_s about z' and then θ_s. Thus the final direction $\vec{p'}$ is given by

$$\vec{p'}/p' = \exp(-i\phi l_z)\exp(-i\theta l_y)\exp(-i\phi_s l_z)\exp(-i\theta_s l_y)z \qquad \text{(C.9)}$$

After some algebra one finds

$$p'_x = |p'|[\cos\phi(\cos\theta\sin\theta_s\cos\phi_s + \cos\theta_s\sin\theta) - \sin\theta_s\sin\phi\sin\phi_s]$$

$$p'_y = |p'|[\sin\phi(\cos\theta\sin\theta_s\cos\phi_s + \cos\theta_s\sin\theta) + \sin\theta_s\cos\phi\sin\phi_s]$$

$$p'_z = |p'|[\cos\theta\cos\theta_s - \sin\theta\sin\theta_s\cos\phi_s]$$

6. We now move back from the CM of the two particles to the original frame. If we came from step 4, the momenta of the particles did not change. If we came from step 5, the momenta changed. Once all the collisions are checked, the particles propagate with $\vec{r_i}(t + \Delta t) = \vec{r_i} + \Delta t(\vec{p_i}/\sqrt{p_i^2 + m_i^2})$.

Appendix D

Wigner Transform and Related Topics

The quantity $f(\vec{r}, \vec{p})$ is a phase space density. It is a classical concept but often arises in semi-classical physics. If $f(\vec{r}, \vec{p})$ is obtained through a Wigner transform of an orbital then the behaviour of the transform can mimic some properties of the orbital. For example, let $\psi(\vec{r})$ be an orbital. The Wigner transform of this is

$$f(\vec{r}, \vec{p}) = \frac{1}{(2\pi\hbar)^3} \int \exp\left(-i\frac{\vec{p}}{\hbar} \cdot \vec{s}\right) \psi\left(\vec{r} + \frac{\vec{s}}{2}\right) \psi^*\left(\vec{r} - \frac{\vec{s}}{2}\right) d^3s \quad \text{(D.1)}$$

The value of $f(\vec{r}, \vec{p})$ can, for some values of $\vec{r}, \vec{p}$, become negative which is obviously unphysical but we will not meet such examples here. We have

$$\int f(\vec{r}, \vec{p}) d^3p = \psi(\vec{r})\psi^*(\vec{r}) = \rho(\vec{r}). \quad \text{(D.2)}$$

Consider a quantum mechanical wave function:

$$\psi_\lambda(\vec{r}) = \left[\frac{\alpha}{\sqrt{\pi}}\right]^{3/2} \exp\left\{-\frac{1}{2}\alpha^2(\vec{r} - \vec{r}_\lambda)^2\right\} \exp\left\{\frac{i\vec{p}_\lambda}{\hbar} \cdot \vec{r}\right\} \quad \text{(D.3)}$$

The Wigner transform phase space density is then

$$f(\vec{r}, \vec{p}) = \frac{1}{(\hbar\pi)^3} \exp\{-\alpha^2(\vec{r} - \vec{r}_\lambda)^2\} \exp\left\{\frac{(\vec{p} - \vec{p}_\lambda)^2)}{\hbar^2\alpha^2}\right\} \quad \text{(D.4)}$$

This obviously mimics some behaviour of the wave function above. An often used equation for $f(\vec{r}, \vec{p})$ in nuclear physics is

$$f(\vec{r}, \vec{p}) = \frac{4}{h^3}\theta(p_F - p)\theta(R - r) \qquad (D.5)$$

One also writes

$$f(\vec{r}, \vec{p}) = \frac{4}{h^3}\theta(p_F(r) - p)$$

Here θ is the Heaviside function. The factor 4 is for spin–isospin degeneracy and h^3 is the "volume" of a quantum state. We have $\rho(\vec{r}) = \frac{4}{h^3}\int\theta(p_F - p)d^3p\,\theta(R - r) = \frac{16\pi}{3h^3}p_F^3$ for $r \le R$. This is a standard formula relating local density to local Fermi surface in Thomas–Fermi theory. For fermions filling up a sphere of radius p_F the average kinetic energy is $\frac{3}{10m}p_F^3 = \frac{3}{10m}(\frac{3h^3}{16\pi})^{2/3}\rho^{2/3}$. With Skyrme parametrisation the potential in which a particle moves is $U(\rho)$. For example we used a potential energy density $v(\rho) = \frac{A}{2}(\frac{\rho}{\rho_0})^2 + \frac{B}{\sigma+1}(\frac{\rho}{\rho_0})^{\sigma+1}$ which gives a potential energy per particle $\frac{A}{2}(\frac{\rho}{\rho_0}) + \frac{B}{\sigma+1}(\frac{\rho}{\rho_0})^{\sigma}$. Adding to this the kinetic energy per particle will allow us to choose the "best" ρ for any nucleus. In this model any deviation from this optimum ρ will raise the energy. So in this model, any nucleus in the ground state has a constant density both inside and near the surface and a sharp surface. This is a feature of Skyrme parametrisation in semi-classical approximation only. This simple feature will disappear in quantum mechanics of course. Also if, in addition to the Skyrme parametrisation, one also has finite range two-body interaction, the sharp surface will have to be replaced by a realistic diffuse surface. In Chapter 12, we introduce a Lenk–Pandharipande method of introducing a diffuse surface.

Appendix E

Thomas–Fermi Model

In 1920's Llewellyn Thomas and Enrico Fermi introduced an approximate semi-classical method for explaining the electron density and ground state energy of atoms [139, 140]. In this method the atom is described as uniformly distributed electron around the nucleus in a six-dimensional phase space and the ground state energy of the atom is expressed as a function of local density. Later on, the Thomas–Fermi method is successfully extended for determining the ground state properties of nuclear matter as well as finite nuclei. The aim of this section is to calculate the ground state energy and corresponding density profile from Thomas–Fermi method for finite nuclei with diffuse surface. Consider a nucleus of A_0 nucleons. The total energy (non-relativistic) can be written as

$$E = \int f(\vec{r}, \vec{p}) \frac{p^2}{2m} d\vec{r} d\vec{p} + \int V(\vec{r}) d\vec{r} \qquad (E.1)$$

where $V(\vec{r})$ is the potential energy density at position $\vec{r}$ and $f(\vec{r}, \vec{p})$ is the phase space density near $(\vec{r}, \vec{p})$. The goal of the Thomas–Fermi method is to minimise this energy subject to $\rho(\vec{r}) = \int f(\vec{r}, \vec{p}) d\vec{p}$ under the condition of total particle number conservation, i.e.

$$\int \rho(\vec{r}) d\vec{r} = A_0 \qquad (E.2)$$

Using Lagrange multiplier (λ), one can then do an unconstrained minimisation of the quantity

$$E' = \int f(\vec{r}, \vec{p}) \frac{p^2}{2m} d\vec{r} d\vec{p} + \int V(\vec{r}) d\vec{r} + \lambda \left(A_0 - \int \rho(\vec{r}) d\vec{r} \right) \quad \text{(E.3)}$$

For the lowest energy state, at each $\vec{r}$, $f(\vec{r}, \vec{p})$ is to be non-zero from 0 to some maximum $p_F(\vec{r})$. Thus,

$$f(r, p) = \frac{4}{h^3} \theta[p_F(r, p) - p] \quad \text{(E.4)}$$

The factor 4 is due to spin–isospin degeneracy and assuming spherical symmetry, one can drop the vector sign on r and p. This leads to

$$E' = \frac{3}{10m} \left\{ \frac{3h^3}{16\pi} \right\}^{2/3} \int \rho(r)^{5/3} d^3 r + \int V(\vec{r}) d\vec{r} + \lambda \left(A_0 - \int \rho(\vec{r}) d\vec{r} \right) \quad \text{(E.5)}$$

Under the variation $\rho(\vec{r})$ to $\rho(\vec{r}) + \delta\rho(\vec{r})$, the change in E' becomes

$$\delta E' = \int d\vec{r} \left[\frac{1}{2m} \left\{ \frac{3h^3}{16\pi} \right\}^{2/3} \rho(r)^{2/3} + U(\vec{r}) - \lambda \right] \delta\rho(\vec{r}) \quad \text{(E.6)}$$

where the potential $U(\vec{r})$ is the functional derivative of the potential energy density with respect to $\rho(\vec{r})$. Since $\delta\rho(\vec{r})$ is arbitrary, Eq. (E.6) will be only zero if at each $\vec{r}$, the following condition holds good.

$$\frac{1}{2m} \left\{ \frac{3h^3}{16\pi} \right\}^{2/3} \rho(r)^{2/3} + U(\vec{r}) - \lambda = 0 \quad \text{(E.7)}$$

This is known as the Thomas–Fermi equation, and by solving this equation ground state density profile and ground state energy can be obtained. To solve Eq. (E.7), one has to start from a guess density. For example, initially we can start with Myers density profile [88], which is given by

$$\rho_{guess}(r) = \rho_M \left[1 - \left[1 + \frac{R}{a} \right] \exp(-R/a) \frac{\sinh(r/a)}{r/a} \right], \quad r < R$$

(E.8)

$$\rho_{guess}(r) = \rho_M [(R/a)\cosh(R/a) - \sinh(R/a)] \frac{e^{-r/a}}{r/a}, \quad r > R$$

(E.9)

where $\rho_M = 1.18 A^{1/3}$ fm and $a = 1/\sqrt{2}$ fm. ρ_M and a determine the equivalent sharp radius and width of the surface respectively.

The ground state density and binding energy calculation from Thomas–Fermi model for two types of interaction potential which are commonly used in intermediate energy nuclear physics are described below.

1. Skyrme + Yukawa Potential: To get diffuse nuclear surfaces in semi-classical calculations, in addition to zero range Skyrme interaction, contribution from finite range Yukawa interaction should be included. Therefore the form of the potential will be

$$U(\vec{r}) = A'' \rho(\vec{r}) + B'' \rho^\sigma(\vec{r}) + \int u_y(\vec{r}, \vec{r}') \rho(\vec{r}') d^3 r' \qquad (E.10)$$

where $u_y(\vec{r}, \vec{r}')$ is the finite range Yukawa potential

$$u_y(\vec{r}, \vec{r}') = V_0 \frac{e^{-|\vec{r}-\vec{r}'|/a}}{|\vec{r}-\vec{r}'|/a} \qquad (E.11)$$

The constants are taken as $A'' = -1563.6\,\text{MeVfm}^3$, $B'' = 2805.3\,\text{MeVfm}^{7/6}$, $\sigma = 7/6$, $V_0 = -668.65\,\text{MeV}$ and $a = 0.45979\,\text{fm}$. Note that, for infinite nuclear matter, the contribution of Yukawa term to the total energy per nucleon reduces to $2\pi V_0 a^3 \rho_0$ but for finite nuclei the Thomas–Fermi solution produces realistic ground state energies and densities.

The Thomas–Fermi equation can be written as

$$\frac{1}{2m}\left\{\frac{3h^3}{16\pi}\right\}^{2/3}\rho(r)^{2/3} + A''\rho(\vec{r}) + B''\rho^\sigma(\vec{r}) = \lambda - \int u_y(\vec{r},\vec{r'})\rho(\vec{r'})d^3r'$$

(E.12)

To solve this equation numerically, r and $\rho_{guess}(r)$ can be discretised into $\{r_i\}$, $\{\rho_{i_{guess}}\}$. Equation (E.7) is solved by using Newton–Raphson method at each i for a particular guess value of $\lambda = \lambda_1$. After iterations, the densities $\{\rho_{1_i}\}$ are obtained which satisfy the condition $\int \rho_1(r)d^3r = A_1$ (say). Similarly for $\lambda_2 = \lambda_1 + \delta$ (δ is a small number) and $\{\rho_{i_{guess}}\}$, one can repeat the earlier steps and another density profile $\rho_2(r)$ can be obtained which satisfies $\int \rho_2(r)d^3r = A_2$ (say). Then a better guess of λ can be obtained from the equation $\lambda_0 = \lambda_1 + \{(A_0 - A_1)/(A_2 - A_1)\}(\lambda_2 - \lambda_1)$. Again by using Newton–Raphson method with inputs as λ_0 and $\rho_{guess}(r)$, $\rho_0(r)$ is determined. If $\int \rho_0(r)d^3r$ is A_0, then the calculation can be stopped, otherwise all the steps (by changing $\rho_{guess}(r) = \rho_0(r)$ and $\lambda_1 = \lambda_0$) are repeated until the required number of particles A_0 is obtained. This method has been used in the past to construct Thomas–Fermi solutions relevant heavy ion collisions [141].

Figure E.1(a) shows the density profile for $A = 124$ nucleus and Fig. E.1.(b) indicates the variation of energy per nucleon with

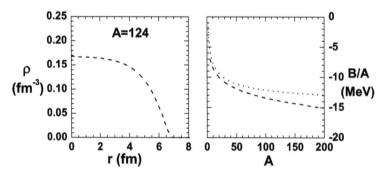

Fig. E.1: Left panel: Density profile for $A = 124$ nucleus obtained by solving Thomas–Fermi equation for interaction potential given in Eq. (E.10). Right panel: Variation of energy per nucleon with mass number obtained from Thomas–Fermi method (dashed line), compared to a typical liquid drop formula $e = -16 + 18A^{-1/3}$ (dotted line).

mass calculated for Thomas–Fermi method with $U(\vec{r})$ given by Eq. (E.10).

2. Skyrme potential with ∇^2 correction: Eq. (E.10) can produce realistic ground state energies and diffuse nuclear surfaces, but for larger finite nuclei the calculation is very time consuming and needs huge computer memory. A ∇^2 correction with the original Skyrme potential can overcome this problem. This form of potential was suggested by Lenk and Pandharipande [98] and is given by

$$U(\vec{r}) = A'\rho(\vec{r}) + B'\rho^\sigma(\vec{r}) + \frac{C}{\rho_0^{2/3}}\nabla_r^2\left[\frac{\rho(\vec{r})}{\rho_0}\right] \qquad (E.13)$$

where $A' = \frac{A}{\rho_0} = -2230.0\,\mathrm{MeVfm}^3$, $B' = \frac{B}{\rho_0^\sigma} = 2577.85\,\mathrm{MeVfm}^{7/6}$ and $\sigma = 7/6$ and the constant in correction term $C = -6.5\,\mathrm{MeV}$. The Thomas–Fermi equation will be

$$\frac{1}{2m}\left\{\frac{3h^3}{16\pi}\right\}^{2/3}\rho(r)^{2/3} + A'\rho(\vec{r}) + B'\rho^\sigma(\vec{r}) + \frac{C}{\rho_0^{2/3}}\nabla_r^2\left[\frac{\rho(\vec{r})}{\rho_0}\right] - \lambda = 0$$

$$(E.14)$$

By substituting $y(r) = r\rho(r)$, Eq. (E.14) becomes

$$\frac{1}{r}\frac{d^2y}{dr^2} + \frac{1}{2mC}\left(\frac{3h^3}{16\pi}\right)^{2/3} + \left(\frac{y}{r}\right)^{2/3} + \frac{A'}{C}\frac{y}{r} + \frac{B'}{C}\left(\frac{y}{r}\right)^\sigma = \frac{\lambda}{C} \qquad (E.15)$$

So, $y(r)$ vanishes both at $r=0$ and $r=\infty$, i.e. Eq. (E.15) becomes a boundary value problem. Therefore similar to the earlier case, in a finite mesh, r and y can be discretized into $\{r_i\}$, $\{y_i\}$, $i = 0, 1, 2, \ldots, M$ ($r_M = Mh$) and the boundary conditions are implemented as $y_0 = 0$ and $y_M = 0$. $\frac{d^2y}{dr^2}$ can be expressed in terms of finite differences

$$\frac{d^2y(r_i)}{dr^2} = \frac{1}{h^2}(y_{i+1} + y_{i-1} - 2y_i) \qquad (E.16)$$

Substituting Eq. (E.16) into Eq. (E.15), leads to a system of non-linear algebraic equations for $\{y_i\}$, which can be solved by multidimensional Newton's method for λ_1. Then by applying the

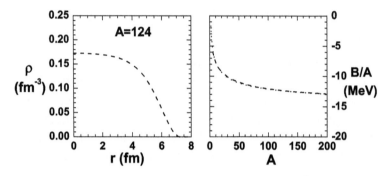

Fig. E.2: Same as Fig. E.1, but with Skyrme interaction, ∇^2 correction term is added (Eq. (E.13)) in the potential for Thomas–Fermi calculation.

same procedure as mentioned in the earlier case, one can determine the ground state density.

Figure E.2(a) shows the density profile for $A = 124$ nucleus and Fig. E.2(b) indicates the variation of energy per nucleon with mass calculated for Thomas–Fermi method with $U(\vec{r})$ given in Eq. (E.13).

Appendix F

Critique of Fluctuation

In Chapter 11 we described how Pauli blocking is examined for test particle collisions. When two tp's collide they change from $(\vec{r_1}, \vec{p_1}), (\vec{r_2}, \vec{p_2})$ to $(\vec{r_1}, \vec{p_1'}), (\vec{r_2}, \vec{p_2'})$, provided fermionic limits are not exceeded if the changes are made. In Chapter 11, volumes in phase space which allow upto 8 tp's centred around $(\vec{r_1}, \vec{p_1'})$ and around $(\vec{r_2}, \vec{p_2'})$ were examined. If the occupancy is less than 8, the transition is allowed and a Monte Carlo decision was taken for the collision to happen.

In the proposed model of fluctuation (Chapter 15) two tp's i and j which passed the test of Pauli blocking were chosen, but when the transition was granted not only these two changed their positions in phase space but $(\tilde{N} - 1)$ closest to tp i are moved. We label these tp's as i_s. Similarly $(\tilde{N} - 1)$ tp's closest to tp j are moved. We label these as j_s. Some of these movements must have violated the fermionic limit and here we try to get an estimate. In accordance with past calculations, choose a volume of phase space where 8 tp's is the maximum number of tp's that can be put in. If n is the number of tp's (not including i_s) already in the unit cell we define an availability factor $\tilde{f} = 1.0 - n/8$. If $\tilde{f} = 0$ we are already at the limit of fermionic occupation. If $\tilde{f}$ is negative we have crossed the quantum limit and are in the classical regime. Any positive number for $\tilde{f}$ between 0 and 1 can accommodate additional fermion. Initially we have two ions in their ground states with correct fermionic occupation. It is also

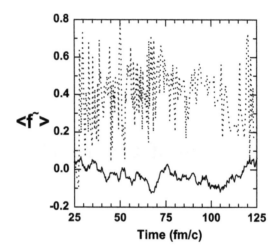

Fig. F.1: Variation of average availability factor (see text) with time (dotted line) for $N_A = 120$ on $N_B = 120$ reaction at beam energy 100 MeV/nucleon. The lower curve (solid line) is the average availability factor $\langle \tilde{f} \rangle$ at the phase space points of arbitrarily chosen 120 test particles in an isolated mass 120 nucleus as they move in time. The fluctuations from the value 0 reflects uncertainties, probably due to fluctuation in initial Monte Carlo simulations.

easy to show that at freeze-out there is no problem with fermionic limits. It is in transition from the initial to the final disassembly that we need to look at. For each collision there are 200 $\tilde{f}$'s to be calculated so we get an average $\tilde{f}$ and plot them in Fig. F.1. We have shown results from $t = 25$ to $125\,\text{fm}/c$ when most of the action takes place. The case shown is for 120 on 120 at $100\,\text{MeV}/n$ beam energy ($50\,\text{MeV}/n$ was studied also). For reference we also plot average $\tilde{f}$ for randomly chosen tp's in a static nucleus of mass 120 in its ground state as they move around in time. This number should ideally be 0 and not fluctuate. The deviations from zero in the static case probably arise due to fluctuations in Monte Carlo sampling. This degree of uncertainty must be also present in the values of $\tilde{f}$ we have plotted in Fig. F.1 for collisions. In spite of these uncertainties the predominantly positive values of $\tilde{f}$ lead us to believe that the general trends we find in our calculations will hold.

If in a collision all of the tp's moved to locations where $\tilde{f}$'s were all positive, we stay within fermionic limits. In case there is a tp which does not satisfy this, one can try to improve the situation by discarding that tp and choosing the next tp to be part of the cloud. Complications arise because when some of the previously chosen tp's are discarded for new ones, the average momentum of the cloud will change, new $\vec{\Delta p}$ will have to be used so the final resting spots obeying energy and momentum conservation will change too. An iterative procedure needs to be formulated but the convergence may be slow.

Alternative methods have been proposed. The two papers which give procedural details of moving two clouds of tp's from initial positions to final positions with a stricter adherence to fermionic limits are [113, 114]. Multiplicity distributions are not given so we cannot compare.

Bibliography

[1] K. Huang, *Statistical Mechanics* (John Wiley and Sons, New York, second edition, 1987), p. 156 and Chapter 14.

[2] F. Reif, *Fundamentals of Statistical and Thermal Physics* (McGraw-Hill, New York, 1965), Chapter 8.

[3] J. Gosset, J. I. Kapusta and G. D. Westfall, *Phys. Rev. C* **18**, 844 (1978).

[4] K. A. Bugaev, M. I. Gorenstein, I. N. Mishustin and W. Greiner, *Phys. Rev. C* **62**, 044320 (2000).

[5] K. A. Bugaev, M. I. Gorenstein, I. N. Mishustin and W. Greiner, *Phys. Lett. B* **498**, 144 (2001).

[6] K. C. Chase and A. Z. Mekjian, *Phys. Rev. C* **52**, 2339 (1995).

[7] A. Bohr, B. Mottleson, *Nuclear Structure*, Vol. I (Benjamin, Reading, MA, 1975).

[8] J. P. Bondorf, A. S. Botvina, A. S. Iljinov, I. N. Mishustin and K. Sneppen, *Phys. Rep.* **257**, 133 (1995).

[9] L. G. Moretto *et al.*, *Phys. Rep.* **287**, 249 (1997).

[10] E. Stanley, *Introduction to Phase Transition and Critical Phenomena* (Oxford Science Publications, 1971).

[11] S. Das Gupta and A. Z. Mekjian, *Phys. Rev. C* **57**, 1361 (1998).

[12] P. Bhattacharyya, S. Das Gupta and A. Z. Mekjian, *Phys. Rev. C* **60**, 054616 (1999).

[13] C. B. Das, S. Das Gupta, W. G. Lynch, A. Z. Mekjian and M. B. Tsang, *Phys. Rep.* **406**, 1 (2005).

[14] S. Das Gupta, *Nucl. Phys. A* **822**, 41 (2009).

[15] P. Bhattacharyya, S. Das Gupta and A. Z. Mekjian, *Phys. Rev. C* **60**, 064625 (1999).

[16] A. S. Botvina *et al.*, *Nucl. Phys. A* **475**, 663 (1987).

[17] K. Sneppen, *Nucl. Phys. A* **470**, 213 (1987).

[18] A. Botvina, G. Chaudhuri, S. Das Gupta and I. Mishustin, *Phys. Lett. B* **668**, 414 (2008).

[19] S. Pratt and S. Das Gupta, *Phys. Rev. C* **62**, 044603 (2000).

[20] L. Shi and S. Das Gupta, *Phys. Rev. C* **70**, 044602 (2004).
[21] L. Beaulieu, L. Phair, L. G. Moretto and G. J. Wozniak, *Phys. Rev. Lett.* **81**, 770 (1998).
[22] L. G. Moretto *et al.*, *Phys. Rev. Lett.* **74**, 1530 (1995).
[23] D. H. S. Gross, *Phys. Rep.* **279**, 119 (1997).
[24] J. Randrup and S. E. Koonin, *Nucl. Phys. A* **474**, 173 (1987).
[25] R. J. Charity, *Nucl. Phys. A* **483**, 371 (1988).
[26] D. Durand, *Nucl. Phys. A* **541**, 266 (1992).
[27] A. Majumder and S. Das Gupta, *Phys. Rev. C* **61**, 034603 (2000).
[28] W. P. Tan *et al.*, *Phys. Rev. C* **68**, 034609 (2003).
[29] G. Chaudhuri and S. Mallik, *Nucl. Phys. A* **849**, 190 (2011).
[30] V. Weiskoff, *Phys. Rev.* **52**, 295 (1937).
[31] J. E. Lynn, *Theory of Neutron Resonance Reactions* (Clarendon, Oxford, 1968), p. 325.
[32] M. B. Tsang *et al.*, *Phys. Rev. C* **64**, 054615 (2001).
[33] G. Chaudhuri, S. Das Gupta and M. Mocko, *Nucl. Phys. A* **813**, 293 (2008).
[34] M. Mocko *et al.*, *Phys. Rev. C* **74**, 054612 (2006).
[35] M. Mocko *et al.*, *Phys. Rev. C* **76**, 014609 (2007).
[36] S. Mallik and G. Chaudhuri, *Phys. Lett. B* **727**, 282 (2013).
[37] S. Mallik and G. Chaudhuri, *Phys. Rev. C* **87** 011602 (2013).
[38] M. Mocko *et al.*, *Phys. Rev. C* **78**, 024612 (2008).
[39] J. Reinhold *et al.*, *Phys. Rev. C* **58**, 247 (1998).
[40] R. Ogul *et al.*, *Phys. Rev. C* **83**, 024608 (2011).
[41] S. Mallik, G. Chaudhuri and S. Das Gupta, *Phys. Rev. C* **83**, 044612 (2011).
[42] M. B. Tsang, *Phys. Rev. Lett.* **71**, 1502 (1993).
[43] S. Mallik, G. Chaudhuri and S. Das Gupta, *Phys. Rev. C* **84**, 054612 (2011).
[44] S. Albergo *et al.*, *Nuove Cimento* **89**, 1 (1985).
[45] M. B. Tsang, W. G. Lynch, H. Xi and W. A. Friedman, *Phys. Rev. Lett.* **78**, 3836 (1997).
[46] M. Mocko, Ph.D. Thesis, Michigan State University (2006).
[47] M. Mocko *et al.*, *Eur. Phys. Lett.* **79** 12001 (2007).
[48] M. B. Tsang *et al.*, *Phys. Rev. C* **76**, 041302 (2007).
[49] G. Chaudhuri *et al.*, *Phys. Rev. C* **76**, 067601 (2007).
[50] G. Chaudhuri, S. Mallik and S. Das Gupta, *Pramana J. Phys.* **82**, 907 (2014).
[51] G. Chaudhuri, S. Mallik and S. Das Gupta, *J. Phys. Conf. Series* **420**, 012098 (2013).
[52] S. Mallik, S. Das Gupta and G. Chaudhuri, *Phys. Rev. C* **89**, 044614 (2014).
[53] J. E. Finn *et al.*, *Phys. Rev. Lett.* **49**, 13219 (1982).
[54] A. S. Hirsch *et al.*, *Phys. Rev. C* **29**, 508 (1984).
[55] C. J. Waddington and P. S. Freier, *Phys. Rev. C* **31**, 888 (1985).

[56] D. Stauffer and A. Aharony, *Introduction to Percolation Theory* (Taylor and Francis, London, 1992).

[57] S. Das Gupta, S. Mallik and G. Chaudhuri, *Phys. Rev. C* **97**, 044605 (2018).

[58] J. Pan and S. Das Gupta, *Phys. Rev. C* **51**, 1384 (1995); J. Pan and S. Das Gupta, *Phys. Lett. B* **344**, 29 (1995).

[59] N. Metropolis, A. W. Rosenbluth, M. N. Rosenbluth, A. H. Teller and E. Teller, *J. Chem. Phys.* **21**, 1087 (1953).

[60] S. Das Gupta and S. K. Samaddar, in *Isospin Physics in Heavy-Ion Collisions at Intermediate Energies*, eds. B.-A. Li and W. Udo Schroder (Nova Science Publishers Inc., Huntington, New York, 2001), p. 109.

[61] X. Campi and H. Krivine, *Nucl. Phys. A* **620**, 46 (1997).

[62] A. Coniglio and W. Klein, *J. Phys. A* **13**, 2775 (1980).

[63] P. Chomaz, F. Gulminelli and V. Duflot, in *Isospin Physics in Heavy-Ion Collisions at Intermediate Energies*, eds. B.-A. Li and W. Udo Schroder (Nova Science Publishers Inc., Huntington, New York, 2001), p. 131.

[64] J. Pan, S. Das Gupta and M. Grant, *Phys. Rev. Lett.* **80**, 1182 (1998).

[65] Y. G. Ma *et al.*, *Phys. Rev. C* **60**, 024607 (1999).

[66] J. Pan and S. Das Gupta, *Phys. Rev. C* **57**, 1839 (1998).

[67] F. Gulminelli and Ph. Chomaz, *Phys. Rev. C* **71**, 054607 (2005).

[68] K. Binder and D. P. Landau, *Phys. Rev. B* **30**, 1477 (1984).

[69] M. Pichon *et al.*, *Nucl. Phys. A* **779**, 267 (2006).

[70] G. Chaudhuri, S. Das Gupta and F. Gulminelli, *Nucl. Phys. A* **815**, 89 (2009).

[71] M. E. Fisher, *Physics* **3**, 255 (1965).

[72] M. L. Gilkes *et al.*, *Phys. Rev. Lett.* **73**, 1590 (1994).

[73] J. B. Elliott *et al.*, *Phys. Lett. B* **381**, 35 (1996).

[74] J. A. Hauger *et al.*, *Phys. Rev. Lett.* **77**, 235 (1996).

[75] J. B. Elliott *et al.*, *Phys. Lett. B* **418**, 34 (1998).

[76] R. P. Scharenberg *et al.*, *Phys. Rev. C* **64**, 054602 (2001).

[77] J. Hufner and D. Mukhopadhyay, *Phys. Lett. B* **173**, 373 (1986).

[78] L. Oddershede, P. Dimon and J. Bohr, *Phys. Rev. Lett.* **71**, 3107 (1993).

[79] J. Pochodzalla *et al.*, *Phys. Rev. Lett.* **75**, 1040 (1995).

[80] C. B. Das, S. Das Gupta and A. Majumder, *Phys. Rev. C* **65**, 034608 (2002).

[81] F. Gulminelli and Ph. Chomaz, *Phys. Rev. Lett.* **82**, 1402 (1999).

[82] S. Mallik, G. Chaudhuri, P. Das and S. Das Gupta, *Phys. Rev. C* **95**, 061601 (2017).

[83] P. Das, S. Mallik and G. Chaudhuri, *Phys. Lett. B* **783**, 364 (2018).

[84] R. Wada *et al.*, *Phys. Rev. C* **99**, 024616 (2019).

[85] B. K. Jennings and S. Das Gupta, *Phys. Rev. C* **62**, 014901 (2000).

[86] C. B. Das, S. Das Gupta and B. K. Jennings, *Phys. Rev. C* **70**, 044611 (2004).

[87] J. Cugnon, T. Mizutani and J. Vandermeulen, *Nucl. Phys. A* **352**, 505 (1981).

[88] W. D. Myers, *Nucl. Phys. A* **296**, 177 (1978).

[89] E. A. Uehling and G. A. Uhlenbeck, *Phys. Rev.* **43**, 552 (1933).

[90] G. F. Bertsch, H. Kruse and S. Das Gupta, *Phys. Rev. C* **97**, 673 (1984).

[91] L. W. Nordeim, *Proc. Roy. Soc. (London) A* **119**, 689 (1928).

[92] C. Gregoire, B. Renaud, F. Sebille, L. Vinet and Y. Rafiray, *Nucl. Phys. A* **465**, 317 (1987).

[93] J. Aichelin and G. F. Bertsch, *Phys. Rev. C* **31**, 1730 (1985).

[94] G. F. Bertsch, *Frontiers of Nuclear Dynamics*, eds. R. Broglia and C. H. Dasso (Plenum Press, 1985), p. 277.

[95] G. F. Bertsch and S. Das Gupta, *Phys. Rep.* **160**, 189 (1988).

[96] Y. X. Zhang *et al.*, *Phys. Rev. C* **97**, 034625 (2018).

[97] R. J. Lenk and V. R. Pandharipande, *Phys. Rev. C* **39**, 2242 (1989).

[98] S. Mallik, G. Chaudhuri and S. Das Gupta, *Phys. Rev. C* **91**, 044614 (2015).

[99] S. Hudan, A. Chbihi *et al.*, *Phys. Rev. C* **67**, 064613 (2003).

[100] H. W. Barz *et al.*, *Nucl. Phys. A* **561**, 466 (1993).

[101] S. J. Lee, H. H. Gan, E. D. Cooper and S. Das Gupta, *Phys. Rev. C* **40**, 2585 (1989).

[102] H. S. Xu *et al.*, *Phys. Rev. Lett.* **85**, 716 (2000).

[103] J. D. Frankland *et al.*, *Nucl. Phys. A* **649**, 940 (2001).

[104] C. B. Das, L. Shi and S. Das Gupta, *Phys. Rev. C* **70**, 064610 (2004).

[105] W. Bauer, G. F. Bertsch and S. Das Gupta, *Phys. Rev. Lett.* **58**, 863 (1987).

[106] J. Gallego *et al.*, *Phys. Rev. C* **44**, 463 (1991).

[107] J. Aichelin and H. Stocker, *Phys. Rev. Lett.* **176**, 14 (1986).

[108] G. E. Beauvais, D. H. Boal and J. C. K. Wong, *Phys. Rev. C* **35**, 545 (1987).

[109] S. Ayik and C. Gregoire, *Nucl. Phys. A* **513**, 187 (1990).

[110] J. Randrup and B. Remaud, *Nucl. Phys. A* **514**, 339 (1990).

[111] Ph. Chomaz, G. F, Burgio and J. Randrup, *Phys. Lett. B* **254**, 340 (1991).

[112] J. Rizzo, Ph. Chomaz and M. Colonna, *Nucl. Phys. A* **806**, 40 (2008).

[113] P. Napolitani and M. Colonna, *Phys. Lett. B* **726**, 382 (2013).

[114] S. Mallik, S. Das Gupta and G. Chaudhuri, *Phys. Rev. C* **93**, 041603 (2016).

[115] J. P. Blaizot, D. Gogny and B. Grammaticos, *Nucl. Phys. A* **265**, 315 (1976).

[116] H. A. Gustafsson *et al.*, *Phys. Rev. Lett.* **52**, 1590 (1984).

[117] K. G. R. Doss *et al.*, *Phys. Rev. Lett.* **57**, 302 (1986).

[118] P. Danielewicz and G. Odyniec, *Phys. Lett. B* **157**, 146 (1985).

[119] G. M. Welke, M. Prakash, T. T. S. Kuo, S. Das Gupta and C. Gale, *Phys. Rev. C* **38**, 2101 (1988).

[120] C. Gale, G. F. Bertsch and S. Das Gupta, *Phys. Rev. C* **35**, 1666 (1987).

[121] R. B. Wiringa, *Phys. Rev. C* **38**, 2967 (1988).

[122] C. Gale *et al.*, *Phys. Rev. C* **41**, 1545 (1990).

[123] J. Zhang, S. Das Gupta and C. Gale, *Phys. Rev. C* **50**, 1617 (1994).

[124] R. Stock, *Phys. Rep.* **135**, 259 (1986).

[125] Y. Kitazoe, M. Sano, H. Toki and S. Nagamiya, *Phys. Lett. B* **166**, 35 (1986).

[126] C. Gale, *Phys. Rev. C* **36**, 2152 (1987).

[127] A. Rittenberg *et al.*, *Rev. Mod. Phys.* **43**, 5114 (1971).

[128] J. Aichelin, A. Rosenhauer, G. Peilert, H. Stocker and W. Greiner, *Phys. Rev. Lett.* **58**, 1926 (1987).

[129] J. Aichelin and H. Stocker, *Phys. Lett. B* **176**, 14 (1986).

[130] A. Vicentini, G. Jacucci and V. R. Pandharipande, *Phys. Rev. C* **31**, 1783 (1985).

[131] J. Aichelin, *Phys. Rep.* **202**, 233 (1991).

[132] R. K. Puri, Ch. Hartnack and J. Aichelin, *Phys. Rev. C* **54**, 28 (1996).

[133] A. Le Fevre *et al.*, *Nucl. Phys. A* **945**, 112 (2016).

[134] A. Ono and H. Horiuchi, *Prog. Part. Nucl. Phys.* **53**, 501 (2004).

[135] A. Ono, H. Horiuchi and T. Maruyama, *Phys. Rev. C* **48**, 2946 (1993).

[136] A. Ono, H. Horiuchi, T. Maruyama and A. Ohnishi, *Phys. Rev. C* **47**, 2652 (1993).

[137] H. Feldmeier, *Nucl. Phys. A* **515**, 147 (1990).

[138] L. H. Thomas, *Proc. Cambridge Philos. Soc.* **23**, 542 (1927).

[139] E. Fermi, *Z. Phys.* **48**, 73 (1928).

[140] J. Gallego, S. Das Gupta, C. Gale, S. J. Lee, C. Pruneau and S. Gilbert, *Phys. Rev. C* **44** 463 (1991).

Lightning Source UK Ltd.
Milton Keynes UK
UKHW020751040919
349163UK00001B/37/P